Studies in Big Data

Volume 87

The series "Studies in Big Data" (SBD) publishes new developments and advances in the various areas of Big Data- quickly and with a high quality. The intent is to cover the theory, research, development, and applications of Big Data, as embedded in the fields of engineering, computer science, physics, economics and life sciences. The books of the series refer to the analysis and understanding of large, complex, and/or distributed data sets generated from recent digital sources coming from sensors or other physical instruments as well as simulations, crowd sourcing, social networks or other internet transactions, such as emails or video click streams and other. The series contains monographs, lecture notes and edited volumes in Big Data spanning the areas of computational intelligence including neural networks, evolutionary computation, soft computing, fuzzy systems, as well as artificial intelligence, data mining, modern statistics and Operations research, as well as self-organizing systems. Of particular value to both the contributors and the readership are the short publication timeframe and the world-wide distribution, which enable both wide and rapid dissemination of research output.

The books of this series are reviewed in a single blind peer review process.

Indexed by zbMATH.

All books published in the series are submitted for consideration in Web of Science.

More information about this series at http://www.springer.com/series/11970

Manjusha Pandey · Siddharth Swarup Rautaray
Editors

Machine Learning: Theoretical Foundations and Practical Applications

Springer

Editors
Manjusha Pandey
School of Computer Engineering
KIIT (Deemed to be University)
Bhubaneswar, Odisha, India

Siddharth Swarup Rautaray
School of Computer Engineering
KIIT (Deemed to be University)
Bhubaneswar, Odisha, India

ISSN 2197-6503 ISSN 2197-6511 (electronic)
Studies in Big Data
ISBN 978-981-33-6517-9 ISBN 978-981-33-6518-6 (eBook)
https://doi.org/10.1007/978-981-33-6518-6

This Springer imprint is published by the registered company Springer Nature Singapore Pte Ltd.
The registered company address is: 152 Beach Road, #21-01/04 Gateway East, Singapore 189721, Singapore

Preface

Data science, big data analytics and machine learning are the need of the hour; it has escalated itself in every sphere of our day-to-day lifestyle starting from smart homes to smart agriculture, automobile sector to educational aids and from automated workplaces to circumspect industries; everywhere, the automation has led towards the requirement of machine learning. Machine learning is defined as the embedded capabilities in the computer systems that provide it the ability to learn and adapt without the requirements of explicit instructions each and every time. The same has been made possible by the use of algorithms and statistical models that are use to analyze and draw inferences from patterns in data. The management of huge amount of data that is continuously generated by the automated systems has resulted into rise to concerns regarding data collection efficiency, data processing, analytics and security along with the mandate of machine learning to further automate the processes. The presented edited book titled *Machine Learning—Theoretical Foundations and Practical Applications* is a work consolidating the chapters submitted and invited chapters presented by invited speakers at the 10th Industry Symposium held during 9–12 January 2020 in conjunction with 16th edition of ICDCIT. As a subset of artificial intelligence machine learning aims to provide computers the ability of independent learning without being explicitly programmed with the ability to take intelligent decisions without human intervention. The stream of research is proceeding towards enabling machines to grow and improve with experiences referred to as learning by machines making them more intelligent. Numerous Advantages of machine learning like usefulness for large-scale data processing, large-scale deployments of machine learning is beneficial for improved speed and accuracy in processing, understanding of nonlinearity in the data and generation of function mapping input to output as in supervised learning providing recommendations for solving classification and regression problems, ensuring better customer profiling and understand of their needs and many more the proposed title aimed to cover the following topics, but not limited to like machine learning and its applications, statistical learning, neural network learning, knowledge acquisition and learning, knowledge-intensive learning, machine learning and information retrieval, machine learning for web navigation and mining, learning through mobile data mining, text and multimedia mining through machine learning, distributed and parallel learning algorithms and applications, feature extraction and classification, theories and models for plausible reasoning, computational learning theory, cognitive modelling, hybrid learning algorithms.

This edited book would be targeting for the technical institutes, analytical industries and analytical research institutes as primary audience, and we hope it would be helpful for the future researchers in the field of machine learning.

Bhubaneswar, India

Dr. Manjusha Pandey
Dr. Siddharth Swarup Rautaray

Contents

Editors and Contributors

About the Editors

Dr. Manjusha Pandey presently working as Associate Professor at the School of Computer Engineering, Kalinga Institute of Industrial Technology, Deemed to be University, Bhubaneswar, Odisha, India. She has teaching and research experience of more than 9 years. She did her doctoral degree from Indian Institute of Information Technology, Allahabad, U.P., India; her research interest includes big data analytics, computer networks, intelligent systems, machine learning and similar innovative areas. Her research contribution includes 04 co-edited proceedings/books which include SIS Springer series, more than 65 research publications in reputed conferences, book chapters and journals indexed in Scopus/SCI/ESCI and with a citation index of 600 as on date. As an organizing chair, she has organized 02 international conferences and has been part of different core committees of other conferences and workshops. She has delivered invited talks in different workshops and conferences.

Dr. Siddharth Swarup Rautaray presently working as Associate Professor at the School of Computer Engineering, Kalinga Institute of Industrial Technology, Deemed to be University, Bhubaneswar, Odisha, India. He has teaching and research experience of more than 9 years. He did his doctoral degree from Indian Institute of Information Technology, Allahabad, U.P., India. His research interest includes big data analytics, image processing, intelligent systems, human–computer interaction and similar innovative areas. His research contribution includes 05 co-edited proceedings/books which include ASIC Springer series, more than 60 research publications in reputed conferences, book chapters and journals indexed in Scopus/SCI/ESCI and with a citation index of 1800 as on date. As an organizing chair, he has organized 05 international conferences (ICCAN 2017, ICCAN 2019, 16th ICDCIT 2020, FICTA 2016, FICTA 2017) and has been part of different core committees of other conferences and workshops. He has delivered invited talks in different workshops and conferences.

Contributors

Rakshit Agrawal School of Computer Engineering, KIIT (Deemed to be University), Bhubaneswar, Odisha, India

Arunkumar Balakrishnan ikval Softwares LLP, Coimbatore, India

Raju Surampudi Bapi School of Computer and Information Sciences, University of Hyderabad, Hyderabad, India;
Cognitive Science Lab, International Institute of Information Technology Hyderabad, Hyderabad, India

Parikshit Barua School of Computer Engineering, KIIT (Deemed to be University), Bhubaneswar, India

Chakravarthy Bhagvati School of Computer and Information Sciences, University of Hyderabad, Hyderabad, India

Vishwanath Bijalwan Institute of Technology Gopeshwar, Gopeshwar, India

Subhashree Darshana School of Computer Engineering, KIIT (Deemed to be University), Bhubaneswar, Odisha, India

Adyasha Dash School of Computer Engineering, KIIT (Deemed to be University), Bhubaneswar, India

Neha Gaud SCSIT DAVV Indore, Indore, India

Harshvardhan GM School of Computer Engineering, KIIT (Deemed to be University), Bhubaneswar, Odisha, India

Mahendra Kumar Gourisaria School of Computer Engineering, KIIT (Deemed to be University), Bhubaneswar, Odisha, India

Krutika Injamuri School of Computer and Information Sciences, University of Hyderabad, Hyderabad, India

Junali Jasmine Jena School of Computer Engineering, KIIT (Deemed to be University), Bhubaneswar, Odisha, India

Sushovan Jena Wipro Limited, WIPRO BHDC, Bhubaneswar, Odisha, India

Ashish Jha NIT Rourkela, Rourkela, Odisha, India

Nooria Karimi School of Computer Engineering, KIIT (Deemed to be University), Bhubaneswar, India

Sai Somanath Komanduri School of Computer and Information Sciences, University of Hyderabad, Hyderabad, India

Sandeepan Mahapatra School of Computer Engineering, KIIT (Deemed to be University), Bhubaneswar, India

Arghya Mazumdar NIT Rourkela, Rourkela, Odisha, India

Lipika Mohanty School of Computer Engineering, KIIT (Deemed to be University), Bhubaneswar, Odisha, India

Manjusha Pandey School of Computer Engineering, KIIT (Deemed to be University), Bhubaneswar, Odisha, India

Siddharth Swarup Rautaray School of Computer Engineering, KIIT (Deemed to be University), Bhubaneswar, Odisha, India

Vijay Bhaskar Semwal MANIT Bhopal, Bhopal, India

Bharat Sharma ZS Associates, Pune, India

What Do RDMs Capture in Brain Responses and Computational Models?

Krutika Injamuri, Sai Somanath Komanduri, Chakravarthy Bhagvati, and Raju Surampudi Bapi

Abstract The question of how the human brain represents information, and if insights from this knowledge help us in formulating better deep learning algorithms, is still open. This paper attempts to address these questions by investigating the brain responses when participants viewed images of objects belonging to a variety of categories. We adopt the representational similarity analysis (RSA) framework wherein the brain responses are projected into a representational dissimilarity (RD) space so that responses from multiple subjects can be assessed on a common basis. We performed RSA of the brain responses of 15 participants on a set of 92 images (from the Algonauts challenge data set) Cichy et al. (The Algonauts project: a platform for communication between the sciences of biological and artificial intelligence 2019 [4]). The results reveal that human brain responses exhibit appropriate qualitative similarities observed in the image space. We then evaluated various deep networks such as VGGnet, AlexNet, and Inception in Siamese configuration with RD-loss function and identified hidden layers that had maximal similarity to human RD matrices (RDM). The best layers then were subjected to further analysis by looking at the image pairs corresponding to high and low similarity values both in the early and late visual areas. Our results point out that while deep neural network (DNN) responses mimicking early visual areas seem to indicate both feature- and category-based similarities, those mimicking the late visual areas are mostly categorical and contextual. It seems that while deep networks are capable of mimicking the similarity distributions observed in human brain responses, it is still not clear what the networks are actually learning while trying to minimize the representational dissimilarity loss function. We suggest the use of alternative dissimilarity measures to possibly redress

K. Injamuri and S. S. Komanduri: Equal Contribution.

K. Injamuri · S. S. Komanduri · C. Bhagvati · R. S. Bapi (✉)
School of Computer and Information Sciences, University of Hyderabad, Hyderabad, India
e-mail: bapiks@yahoo.co.in

R. S. Bapi
Cognitive Science Lab, International Institute of Information Technology Hyderabad, Hyderabad, India

M. Pandey and S. S. Rautaray (eds.), *Machine Learning: Theoretical Foundations and Practical Applications*, Studies in Big Data 87,
https://doi.org/10.1007/978-981-33-6518-6_1

this problem. We also discuss the possibility of adopting a dissimilarity metric for training future deep networks.

Keywords fMRI · MEG · RSA · RDM · brain decoding · information representation in the brain

1 Introduction

The central focus of this paper is understanding the representation that the brain uses for storing information, in particular, visual information. One way to analyse this is to use non-invasive methods such as functional magnetic resonance imaging (fMRI), magnetoencephelography (MEG) to record the brain responses when human subjects are either actively or passively processing visual information. The entire set of experiments described in the paper tries to answer the following question: what does the human mind store for visual processing? Moreover, can we have a deep neural network that can mimic the working of the human visual cortex?

Traditional methods such as multi-voxel pattern analysis (MVPA) have been used to identify a set of voxels that are selectively active during the processing of visual information, and these patterns have been compared with those learned by suitably configured deep neural networks [18]. The more recent approach is to perform a representational similarity analysis (RSA) [8]. RSA is performed by calculating the similarity of responses to pairs of images, thus projecting the original data into a representational similarity space. This similarity space can then be used to compare different participants' responses or compare the responses of a human participant with those of a deep neural network. In addition, RSA allows us to go beyond issues such as signal noise and individual variability in responses often encountered in fMRI experiments. We make use of the fMRI recordings of the 92 image set collected by Cichy et al. [3], for all our experiments.

In this study, we discuss the usefulness of RSA to address the questions we posed earlier. We also present the results and insights obtained from our computational experiments on neural nets to see how well they mimic the human responses and conclude by offering suggestions with regard to the usage of RSA.

2 Methods

2.1 Representational Similarity Analysis (RSA)

RSA allows us to compare representations obtained from different modalities (e.g. deep neural networks and fMRI patterns) by characterizing the dissimilarity between the response patterns on a set of stimuli. We use a representational dissimilarity

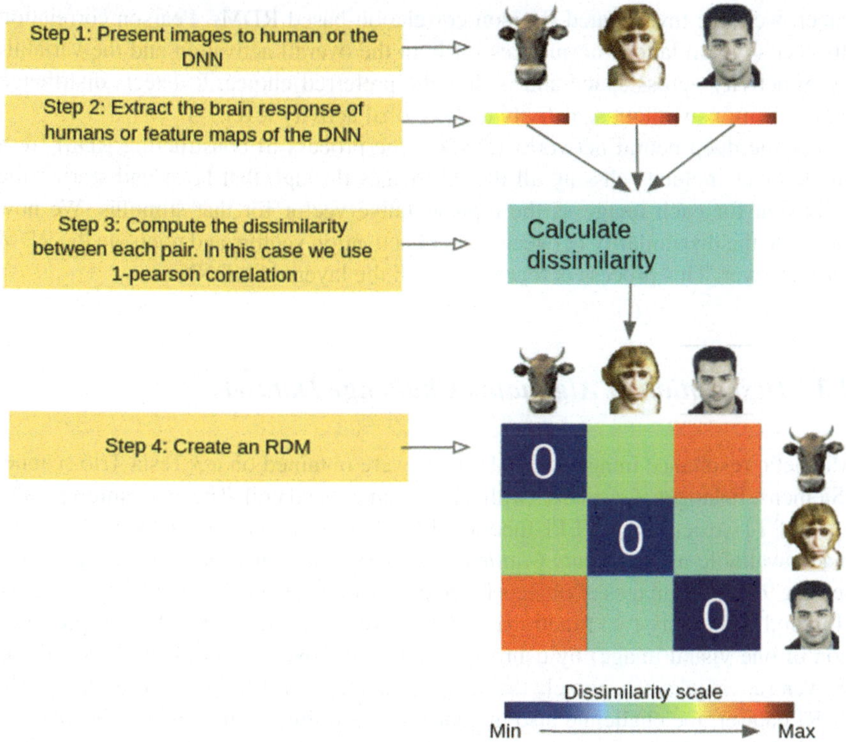

Fig. 1 For each pair of experimental conditions, the associated activity patterns (in a brain region or model) are compared. The dissimilarity between them is measured as one minus the correlation. Dissimilarity is 0 for perfect correlation, 1 for no correlation, 2 for perfect anti-correlation. The dissimilarities for all pairs of conditions are assembled in the RDM. Each cell of the RDM, thus, compares the response patterns of two images

matrix (RDM) to relate different modalities. RDM is a square symmetric matrix in which the diagonal entries are the comparisons between identical stimuli and are 0 by definition (See Fig. 1). Each non-diagonal value indicates the dissimilarity between the response patterns associated with two different stimuli.

2.2 *RDM Construction for Human Brain Responses and Deep Neural Nets (DNNs)*

For a given brain region, we consider the voxel intensities and represent them as an activity vector. By comparing these activity vectors, we obtain the representational dissimilarity matrix (RDM) for human brain responses. Dissimilarity measures such as Pearson correlation, cosine similarity, Gaussian kernels can be used. In the current paper, we have investigated Pearson correlation-based RDMs. Pearson correlation distance (1-correlation) normalizes for both the overall activation and the variability of activity across space and is thus the preferred choice. It detects distributed representations without sensitivity to the global activity level.

For the deep neural networks (DNNs), the process of constructing RDM for a single-layer includes passing all the 92 images through that layer and storing the activation for each image as the representative vector for that stimulus. We now calculate the dissimilarity between pairs of activation vectors and populate the RDM for that layer. This process is repeated for all the layers of a DNN.

2.3 Description of Algonauts Challenge Datasets

Magnetic resonance imaging (MRI) scans were obtained on a 3 Tesla Trio scanner (Siemens, Erlangen, Germany) with a 32-channel head coil. The acquisition parameters of T1-structural and EPI-functional MRI are described in Cichy et al. [3] and made available as *Algonauts Challenge dataset*. The experiment consisted of presenting 92 visual images of objects for passive viewing to each of the 15 participants. Hundred and ninety-two functional volumes were acquired for each trial (presentation of one visual image) by using a partial brain coverage protocol covering only the ventral visual areas, namely the occipital and temporal cortices. In addition to the fMRI dataset, the challenge also includes data acquired from magnetoencephalography (MEG) with a similar experimental protocol (please see [3] for more details).

The Algonauts challenge [3] concentrates on two regions of interest (ROIs) in ventral visual stream: the occipital cortex, denoted here as the *early visual cortex* (EVC) and the late visual activation in the inferior temporal cortex (IT). The neurons in EVC are said to respond to simple features such as edges, and the neurons in IT are said to be responsible for categorical perception of objects [5]. Visual information is processed in the human brain in a hierarchical fashion starting from feature extraction in the EVC and continuing on to object recognition in the IT [5].

The aim of the Algonauts challenge is to construct suitable networks that can mimic the activation in the early and late visual areas from both fMRI and MEG modalities [8]. To accomplish this, RDMs from the fMRI responses of EVC and IT are provided. Similarly, RDMs from responses obtained from MEG corresponding to visual processing in early and late time intervals are also considered in this study.

The training dataset consists of RDMs that were constructed for the said signal space in response to two sets of images, one consisting of 92 images, mostly featuring silhouette objects and the other consisting of 118 images, featuring images with a natural background [3]. The RDMs were also constructed for another set consisting of 78 images held back as the test dataset [8].

2.4 Experiments with DNNs

In line with our goal to establish a deep neural network (DNN) which can mimic the human responses, we fine-tune DNNs with human RDMs. We consider four different DNN architectures for our trials. We use AlexNet [9], two variants of VGG [16], and Inception [17]. We obtain implementations of each of the mentioned architectures from the Pytorch model zoo [12], trained on the ILSVRC12 dataset [14]. Much of our experimental methods were inspired from [1].

The Algonauts challenge provides with two image sets as mentioned earlier, one with 92 images and the other with 118 images. They also provide RDMs associated with each of these image sets. All our experiments use only the 92 image set. The 92 image set has two RDMs, one corresponding to the EVC (early visual cortex) region and the other corresponding to the IT (interior temporal) region. As a consequence of the image set containing 92 images, the RDM of a single participant, for a particular region of interest (EVC or IT), would be of shape 92×92. We therefore have an RDM of shape $15 \times 92 \times 92$, for each of EVC and IT region. We normalize each of the RDMs using z-score [function of *sklearn* toolkit], after which the RDMs were averaged across the first dimension. As a result, we now have an subject-averaged RDM of shape 92×92 for each of the EVC and IT region.

To mimic the human responses using DNN, we take each of the pre-trained architecture and arrange them in Siamese architecture [2]. A Siamese network consists of two sister networks that are identical and share the same weights, which join at their outputs. The output of the network is fed to a loss function, which calculates the distances between the two outputs of the network. These type of networks are really useful to learn the similarities between the inputs. In our case, we want the distance between the outputs of two images to be equal to the corresponding RDM value of the image pair in the empirical data. Since we have 92 images, we will have a total of ${}^{92}C_2$ pairs that constituent the training set. Therefore, our training process includes arranging one of the above-mentioned architectures that was pre-trained on ImageNet, in a Siamese fashion and pass it two inputs I_1 and I_2. If our network is represented by a function G, the corresponding outputs will be $G(I_1)$ and $G(I_2)$, for which we compute $[1 - P(G(I_1), G(I_2))]$ where P is Pearson correlation. Now, the loss is calculated as the squared Euclidean distance between the estimated (as described above) and the observed (in empirical data) correlation between the two input images. The observed value is the same as the average RDM described earlier.

The above procedure is used to fine-tune all the architectures for both EVC and IT averaged RDMs. Furthermore, while fine-tuning the DNNs, we trained it in two configurations. In the first configuration, the weights of all the layers were allowed to be updated during a backward pass. In the second configuration, if it were an EVC fine-tuning, we would freeze the weights of all the layers after 40th one in the Inception model, and the all the layers after the 4th layer in AlexNet and VGG variants; that is, the weights of those layers are not updated. If it were an IT fine-tuning, we would freeze the weights of all the layers before 40th one in the Inception model, and the all the layers before the 4th layer in AlexNet and VGG variants.

Since all the 92 images have no background, we applied Foveation before the images were presented to the models. *Foveation* is an image processing technique that varies the amount of detail depending on the fixation point [13]. The fixation point of the image has the highest resolution. In all our experiments, we used the centre of the image as the only fixation point and used a Gaussian blur that increases as the radius from the fixation point increases. The idea behind this was to emulate the working of the human retina. To understand to what extent Foveation affects the model's correlation with the human RDMs, we presented the models with both foveated and non-foveated images.

For an easy representation of results, the models will have their configuration as subscript. We use 'e' for early and 'l' for late layers. A subscript of 'f' represents that the images were Foveated before shown to the model and the subscript 'n' represents that no Foveation was applied. Subscripts 'p' and 'q' represent frozen and unfrozen configurations, respectively. Therefore, a model named $VGG16_{lfp}$ would be the VGG16 architecture, fine-tuned for the late RDM values (IT region), with no Foveation applied when the image pairs were presented and the initial layers frozen during the fine-tuning process. This training setup gives us a total of 24 models to investigate.

Each of the above models was trained for 150 epochs except for inception, which was trained for 100 epochs. All the models were trained with a learning rate of 0.05. *Adam optimiser* [7] was used for the inception-based models while all others use *Stochastic Gradient Descent*(SGD) optimizer. We used a batch size of 32 for all the models.

2.5 Selecting the Best Models

RSA plays an essential role in the selection process as it facilitates the comparison of model outputs at every layer with those of human fMRI data, belonging to different modalities. Once the RDMs for the EVC and IT regions are constructed for all the layers of the 24 different model configurations we trained, they are now directly comparable with those of the humans. We use *Spearman correlation R* to calculate the similarity and relate the RDMs. The result is then squared to R^2. This depicts the amount of explainable variance. *Noise ceiling* (NC) is the expected RDM correlation achieved by the ideal model, given the noise in the data (see [8] for details). The subject-averaged RDM of the humans is assumed to be the best estimate of the ideal model. The noise ceiling is used to normalize R^2 values to noise-normalized variance explained. Thus, a model can explain from 0 to 100% of the explainable variance. This illustrates the model's explanatory power of the brain data. The model that has the highest correlation for the EVC or IT region would then be said to have the best fit for the responses from the human brain. We ran this test for all the layers of the 24 model configurations to identify the best matching layers. Since we used the correlation score as the only criterion to select the best matching layers, it may be possible that the best matching layers for EVC and IT may come from two different model configurations.

2.6 *Qualitative Analysis of High and Low Values in RDMs*

As per the definition of RD measure, if the RDM value for a cell is low, this means that the corresponding image pairs are highly correlated, and similarly if the RDM value is high, the image pairs would be highly dissimilar. To check if this is indeed the case, we took the images pairs that corresponding to high and low RDM values from both the model RDMs and human response RDMs and performed qualitative analysis on them. For all the 15 participants, we analysed the top 20 similar and dissimilar image pairs selected from the human brain response RDMs. Similarly, we considered the top 20 similar and dissimilar image pairs from the model responses. We looked at all the layers of models from each architecture that showed the highest correlation on average for the EVC and IT regions to identify any consistent behaviour. For example, for VGG19 that has 44 layers, we identified the top 20 similar and dissimilar images for qualitative analysis.

To understand the extent of categorical grouping by the human brain in the EVC and IT regions and also at different layers of the DNNs, we constructed new RDMs, hereafter referred to as *Categorical RDMs*. To construct these categorical RDMs, we considered eight categories used by the fMRI track winner of the Algonauts challenge [10]. The original RDMs provided by the Algonauts are rearranged based on the categories in the following order: scenes-with-objects, fruits–vegetables, animals, animal-face, monkey-face, face, body-parts, and back-profile-view-of-humans. An 8×8 categorical RDM is constructed by averaging the values of the original 92x92 RDM for the images that belong to the same category. This approach of constructing categorical RDMs is inspired from [6] where each cell represents the within-category or between-category distance. Figures 4, 5 and 6 represent the visualization of these RDMs. Blue colour in these figures represents a low RD value, and red colour depicts a high RD value.

3 Results

3.1 *Analysis of the Human Brain Responses*

To investigate what RDMs capture, we averaged the human EVC and IT RDMs of the 92 image set, across all the participants. Then, we selected the top 20 similar image pairs (pairs that have low RD values) and the top 20 dissimilar image pairs (pairs that have high RD values) and few representative examples are displayed in

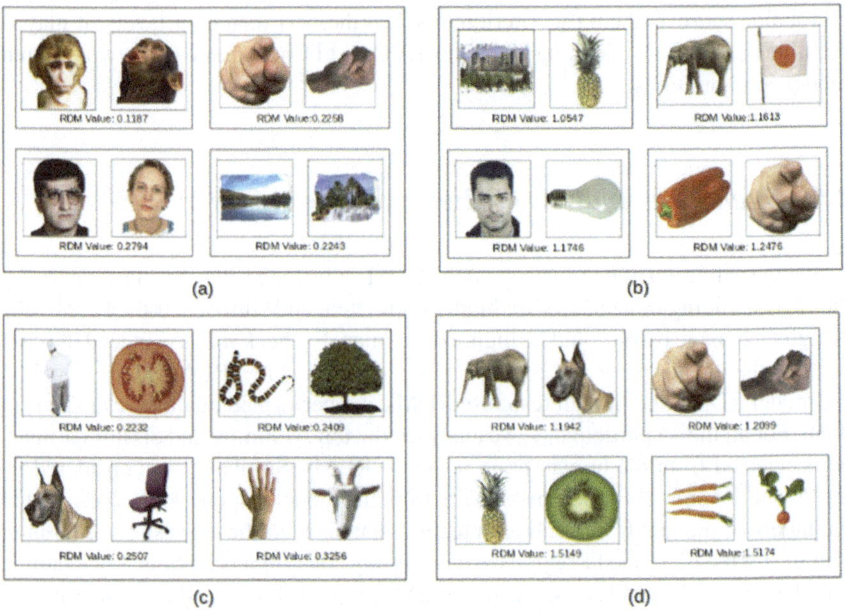

Fig. 2 Similar and dissimilar image pairs in EVC region of averaged human brain responses and their RD values. a Image pairs that have low RD values and are similar. **b** Image pairs that have high RD values and are dissimilar. **c** Image pairs that have low RD values but are dissimilar. **d** Image pairs that have high RD values but are similar

the following figures. Figures 2 and 3 correspond to the good and bad examples of similar and dissimilar image pairs that are selected from the averaged human RDMs of EVC and IT region of interests, respectively.

3.2 Deep Networks Mimicking Human Brain Responses

Table 1 manifests the results of best performing models in EVC and IT in each of the three experiments. The first experiment being the evaluation of pre-trained models (Alexnet, VGG16, VGG19, and Inception) with the ground truth human RDMs is provided by the Algonauts challenge. These results in 1 have the value in fine-tuned column as "No". The second experiment is the evaluation of fine-tuned models with the ground truth and the best results for EVC and IT ROIs on 92 image set. In Table 1 ,these results have the value in fine-tuned column as "Yes" and the image set column as "92". The third experiment is evaluation of these fine-tuned models with the 78 image challenge test set [11]. For this, the value in fine-tuned column is "Yes" and image set is "78".

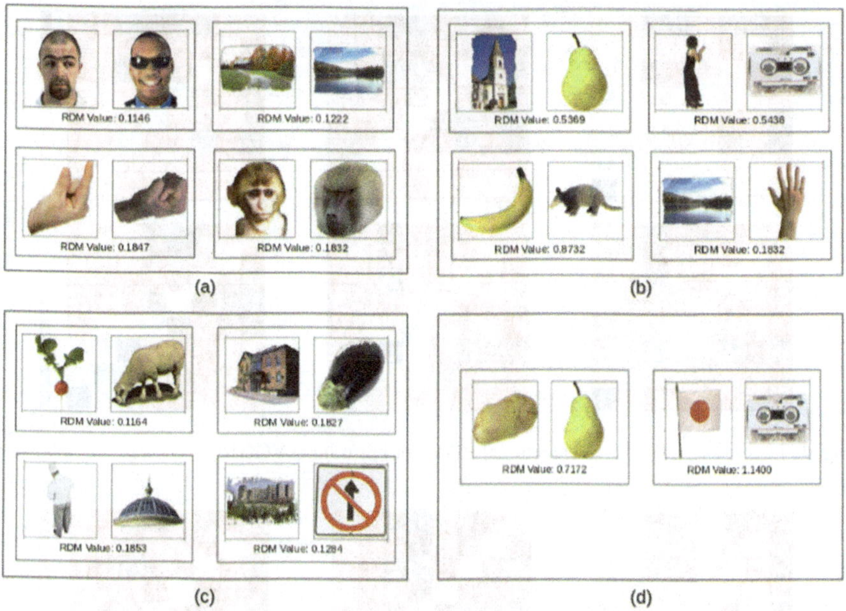

Fig. 3 Similar and dissimilar image pairs in IT region of averaged human brain responses and their RD valcs. **a** Imagc pairs that have low RD values and are similar. **b** Image pairs that have high RD values and are dissimilar. **c** Image pairs that have low RD values but are dissimilar. **d** Image pairs that have high RD values but are similar

Table 1 fMRI track results of ImageNet pre-trained models that performed the best when presented with 92 image set

Model	Fine-tuned	Image set	ROI	Layer name	NC %
Alexnet	No	92	EVC	6: MaxPool2d	30.82
VGG19	No	92	IT	31: Conv2d	13.52
$VGG19_{efq}$	Yes	92	EVC	44: Linear	42.64
$VGG19_{lnp}$	Yes	92	IT	40: Dropout	58.50
$VGG16_{efp}$	Yes	78	EVC	29: Conv2d	20.97
$Inception_{lnq}$	Ycs	78	IT	6: Conv2d_4a_3 × 3	21.30

3.2.1 What Did the Fine-Tuned Models Learn?

To see what the fine-tuned models learned across the layers, we compare and contrast the RDMs at different layers of Imagenet pre-trained VGG19 (Fig. 4) and the fine-tuned $VGG19_{lnp}$ model which performed the best in the IT for the 92 image set. In order to do this, we consider the categorical RDMs (refer to the methods section for more details).

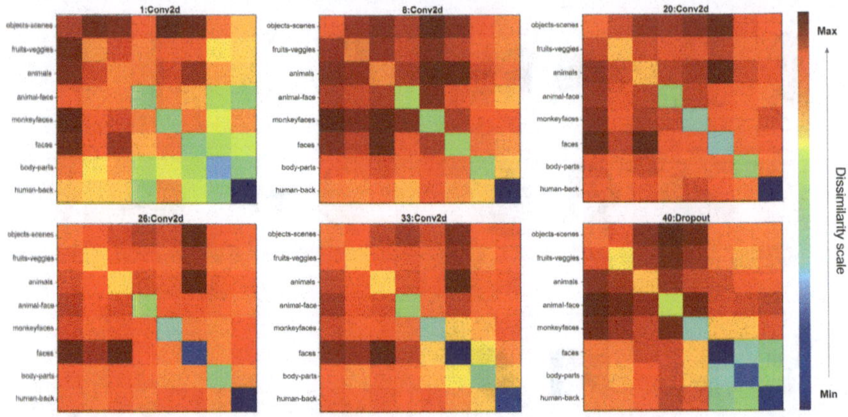

Fig. 4 Layer-wise progress of categorical RDMs of pre-trained VGG19

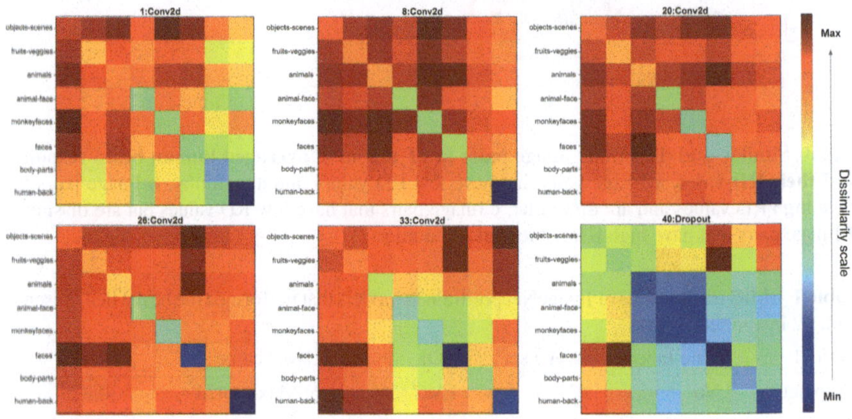

Fig. 5 Layer-wise progress of categorical RDMs of fine-tuned $VGG19_{lnp}$

3.2.2 Similar and Dissimilar Image Pairs

We investigated the RDMs of the models similar to those of the human RDMs. For this purpose, we considered the 44th layer of $VGG19_{efq}$ which highly correlates (42.64% NC) with the EVC ROI of averaged human RDM and the 40th layer of $VGG19_{lnp}$ which highly correlates (58.50 % NC) with the IT ROI of averaged human RDM. Figures 7-I and 7-II depict good and bad image pairs of the best matching models for mimicking EVC and IT activation.

Figure 6 makes a head-to-head comparison of the categorical RDMs of human EVC and IT and the corresponding highly correlated layers of the fine-tuned models.

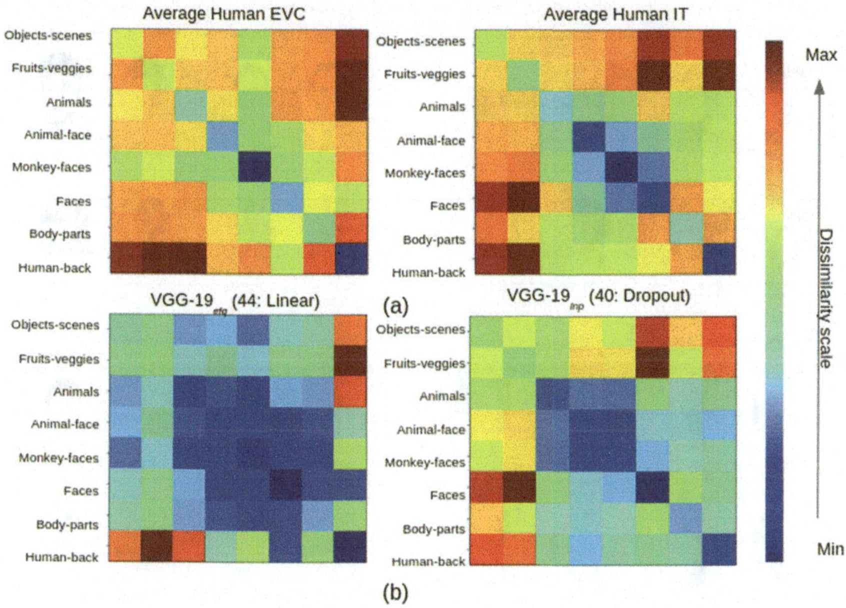

Fig. 6 Comparing the EVC and IT RDMs of human brain response with the highly correlated early and late layers of the fine-tuned model. a displays the average categorical RDM of the EVC and IT of the RDMs averaged across all subjects. **b** depicts the categorical RDM of 44th layer of $VGG19_{efq}$ and 40th layer of $VGG19_{lnp}$ that highly correlated with the human EVC and IT RDMs

4 Discussion

4.1 Qualitative Analysis of Human Responses and DNN Responses

When we collected the images pairs that have low RDM values in the EVC region, we noticed that there were examples of not only similar images but also dissimilar ones. Consider Fig. 2a, we see that human faces and monkey faces are found similar and several such appropriate pairs are shown in this figure. In Fig. 2c, we see groupings that have low RD values. The image pair of a dog and a chair seems to have shape similarities, while other similar pairs like that of a chef and a tomato, a snake and a tree seem to indicate context-dependent grouping. We also identified pairs that are similar and yet have high RD values. In Fig. 2d, we notice examples from different categories such as animals, body parts, and fruits and vegetables that seemed to match in their visual category but the RDM scores indicate otherwise, whereas those pairs in Fig. 2b that have high RD values and are dissimilar seem to be appropriate. This last result where two quite distinctly different objects ending up with very high RD values is a consistent behaviour observed in all the 15 participants.

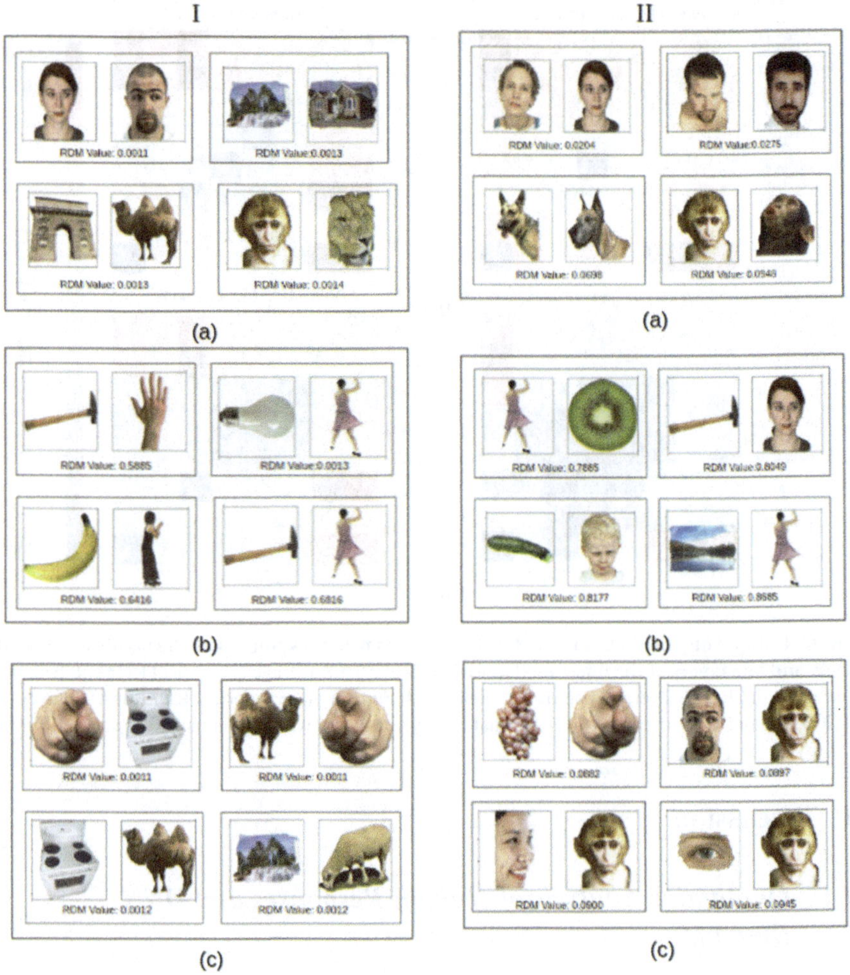

Fig. 7 **(I) Similar and dissimilar image pairs for VGG19 $_{\text{efq}}$ model which is highly correlated with EVC region of average human RDM.** **a** Image pairs that have low RD values and are similar. **b** Image pairs that have high RD values and are dissimilar. **c** Image pairs that have low RD values and are dissimilar. **(II) Similar and dissimilar image pairs for VGG19 $_{\text{lnp}}$ model which is highly correlated with the IT region of average RDM from human responses.** **a** Image pairs that have low RD values and are similar. **b** Image pairs that have high RD values and are dissimilar. **c** Image pairs that have low RD values but are dissimilar

We observe the same trend when we analysed the image pairs for the IT region. Figure 3a shows that objects belonging to the same category are paired together. Figure 3b shows objects that are not related to each other and in line with this the RD values are also high. Figure 3c shows image pairs that are visually dissimilar, yet have low RD values. For example, consider the pair: a vegetable (aubergine/brinjal) and a building are matched with a highly similar brain responses in the IT region! Fig. 3d shows images that belong to the same apparent visual category, but the RD values being high are inappropriate. The overall observation from the qualitative analysis of RDMs from human responses is that the early responses seem to match based on feature and category similarities, and those in the IT regions seem to indicate categorical similarity.

When DNNs are configured to learn these similarity profiles using RD-loss function, it appears that the networks seem to learn progressively more categorical representation as observed in the categorical RDMs of two network configurations in Figs. 4 and 5. This is expected from deep architectures that learn very well to categorize objects.

We did a similar analysis for DNNs. We qualitatively compared the most similar and dissimilar image pairs of the layer that best matched the human EVC and IT RDMs. We tried to guess the reasons behind their pairings and understand what the networks seem to use for making these decisions. Figure 4a shows image pairs that are similar and have low RD values. We observe that human faces seem to be grouped quite often as similar. We also observe that shape and colour are used as a deciding factor. Examples like the monkey and a lion face paired together support this claim. Figure 4b shows object are dissimilar and have high RD values. This behaviour where dissimilar objects have high RD values seems consistent with humans. Figure 4c shows some inconsistent examples. When we compared the image pairs of the layer that best matched the IT RDM, we show similar behaviour. The same trends repeat. In both the best matching layers, we could not find many examples that have high RD values but are actually similar.

Thus, the qualitative analysis of DNN responses indicates that the layers that match early visual responses (EVC in Fig. 7-I) seem to match both feature-wise and category-wise similarities, whereas the responses in the layers that match late visual responses (IT in Fig. 7-II) seem to be more categorical and highly contextual in their representation.

Categorical RDMs of Humans and DNNs: From the categorical RDMs across human and DNN responses as depicted in Fig. 6, we observe that the categorical RDM of 40th layer of $VGG19_{lnp}$ is quite close to categorical RDM of human IT. We also observe the hierarchical grouping that the model learns across the animal-category as indicated by a blue square (compare the top and bottom-left panels in Fig. 6). It needs to be mentioned that models seem to distinguish between animate and inanimate categories (the distinct blue squares on the diagonal in both the bottom-left and right panels of Fig. 6). The categorical RDMs of DNNs (lower panels of Fig. 6) are not as sharp as those of the human response RDMs (especially, the EVC categorical RDM in the top-left panel of Fig. 6). This might be because of the similarity across low-level features such as shape, colour that are present across different images belonging to different categories.

4.2 The Way Forward

Overall, it appears that the representational dissimilarity (RD) measure seems to bring out similarities and dissimilarities in the human responses reasonably well. However, the results of the deep neural networks (DNNs) mimicking the human responses using RD-loss function seem to be a mixed bag. The RDMs computed from DNNs aggregated over eight categories such as animals, faces, humans seem to be noisy when compared to the RDMs from human responses. One possible future direction is to try other correlation measures such as Spearman or Kernel similarity metrics as proposed in Shahbazi, Raizada, and Edelman [15]. It needs to be pointed out that human information representation seems to be highly distributed and very contextual in the sense that an item is stored along with the environmental and internal multi-modal context. It is not clear if RD measure, being linear, can capture such contextual representation. Do the off-diagonal elements in RDMs point to such contextual information, something that needs to be investigated in future? One limitation of the current study is that responses from only two ROIs (EVC and IT) have been considered and the distributed information in other ROIs is not considered. The final, tantalizing proposal is to consider using dissimilarity as a process for constructing deep networks for classification.

Conflict of interest The author(s) declare that they have no conflict of interest.

References

1. Agrawal, A. (2019). Dissimilarity learning via siamese network predicts brain imaging data (2019). arXiv: 1907.02591 [q-bio.NC].
2. Bromley, J., Guyon, I., LeCun, Y., Säckinger, E., & Shah, R. (1994). Signature verification using a "siamese" time delay neural network. In *Advances in Neural Information Processing Systems* (pp. 737–744).
3. Cichy, R. M., Pantazis, D., & Oliva, A. (2016). Similarity-based fusion of meg and fmri reveals spatio-temporal dynamics in human cortex during visual object recognition. *Cerebral Cortex, 26*(8), 3563–3579.
4. Cichy, R. M., Roig, G., Andonian, A., Dwivedi, K., Lahner, B., Lascelles, A., et al. (2019). The algonauts project: A platform for communication between the sciences of biological and artificial intelligence. CoRR http://arxiv.org/abs/1905.05675.
5. Goodale, M. A., & Milner, A. D. (1992). Separate visual pathways for perception and action. *Trends in Neurosciences, 15*(1), 20–25.
6. King, M. L., Groen, I. I., Steel, A., Kravitz, D. J., & Baker, C. I. (2019). Similarity judgments and cortical visual responses reflect different properties of object and scene categories in naturalistic images. *NeuroImage, 197*, 368–382.

7. Kingma, D. P., & Ba, J. (2014). Adam: A method for stochastic optimization. arXiv preprint arXiv: 1412.6980.
8. Kriegeskorte, N., Mur, M., & Bandettini, P. A. (2008). Representational similarity analysis-connecting the branches of systems neuroscience. *Frontiers in Systems Neuroscience*, *2*, 4.
9. Krizhevsky, A., Sutskever, I., & Hinton, G. E. (2012). Imagenet classification with deep convolutional neural networks. In: *Advances in Neural Information Processing Systems* (pp. 1097–1105).
10. Lage-Castellanos, A., & De Martino, F. (2019). Predicting stimulus representations in the visual cortex using computational principles. bioRxiv p. 687731.
11. Mohsenzadeh, Y., Mullin, C., Lahner, B., Cichy, R. M., & Oliva, A. (2019). Reliability and generalizability of similarity-based fusion of meg and fmri data in human ventral and dorsal visual streams. *Vision*, *3*(1), 8.
12. Paszke, A., Gross, S., Massa, F., Lerer, A., Bradbury, J., Chanan, G., et al. (2019). Pytorch: An imperative style, high-performance deep learning library. In: H. Wallach, H. Larochelle, A. Beygelzimer, F. d' Alché-Buc, E. Fox, R. Garnett, (Eds.), *Advances in Neural Information Processing Systems* (Vol. 32, pp. 8024–8035). Curran Associates, Inc. (2019). http://papers.neurips.cc/paper/9015-pytorch-an-imperative-style-high-performance-deep-learning-library.pdf .
13. Perry, J. S., & Geisler, W. S. (2002). Gaze-contingent real-time simulation of arbitrary visual fields. In *Human Vision and Electronic Imaging VII* (Vol. 4662, pp. 57–69). International Society for Optics and Photonics.
14. Russakovsky, O., Deng, J., Su, H., Krause, J., Satheesh, S., Ma, S., et al. (2015). Imagenet large scale visual recognition challenge. *International Journal of Computer Vision*, *115*(3), 211–252.
15. Shahbazi, R., Raizada, R., & Edelman, S. (2016). Similarity, kernels, and the fundamental constraints on cognition. *Journal of Mathematical Psychology*, *70*, 21–34.
16. Simonyan, K., & Zisserman, A. (2014). Very deep convolutional networks for large-scale image recognition. arXiv preprint arXiv: 1409.1556.
17. Szegedy, C., Vanhoucke, V., Ioffe, S., Shlens, J., & Wojna, Z. (2016). Rethinking the inception architecture for computer vision. *Proceedings of the IEEE Conference on Computer Vision and Pattern Recognition*, 2818–2826.
18. Wang, X., Liang, X., Jiang, Z., Nguchu, B.A., Zhou, Y., & Wang, Y: Decoding and mapping task states of the human brain via deep learning. arXiv (2020). https://arxiv.org/abs/1801.09858.

Challenges and Solutions in Developing Convolutional Neural Networks and Long Short-Term Memory Networks for Industry Problems

Arunkumar Balakrishnan

Abstract Developing a deep learning application requires unique approaches and methods. These are discussed in two industrial applications. The architectures used are two classic ones: convolutional neural network (CNN) and long short-term memory (LSTM). The CNN application to identify hand-written choice options from a scanned image, faced issues of "pose" (oriental transformation) problem and "region segmentation." The LSTM application to predict the equated monthly instalment by all customers had to overcome challenges of "multi-variate" and "multiple time series." The evolution of a satisfactory solution for the customer in both these cases is described in this chapter. The importance of data analysis, data pre-processing and proper choice of hyper-parameters is seen through these experiments.

Keywords Convolutional neural network (CNN) · Long short-term memory (LSTM) · Oriental transformation · Region segmentation

1 Introduction

Artificial neural networks try to emulate the behavior of the central nervous system of a human being, through a software system. The human brain consists of trillions of neurons, highly interconnected. Signals are passed from one neuron to many other neurons. Based on the total signal strength received by a neuron and its synapse level (internal threshold setting), the neuron generates an output signal. This simple architecture replicated across trillions of neurons forms the basis for human intelligence. In a similar manner in an artificial neural network, a neuron takes input from a set of other neurons, multiplies each input by a value representing weightage given to the input source and then sums them up. It then compares this summation with its threshold value. If it is greater than the threshold value, then an output is sent to all artificial neurons connected to it. This activity can repeat across multiple layers of

A. Balakrishnan (✉)
ikval Softwares LLP, Coimbatore, India
e-mail: arun@ikval.com

M. Pandey and S. S. Rautaray (eds.), *Machine Learning: Theoretical Foundations and Practical Applications*, Studies in Big Data 87,
https://doi.org/10.1007/978-981-33-6518-6_2

neurons. If the final output matches the expected output, then the network architecture is maintained. Else, if there is an error between expected output and received output, then a learning phase is enabled. In the case of a supervised learning (where the correct output is made available), the difference between the expected output and received output is differentiated with respect to the weightage values for each input connection and this value is used to change the weight values for each connection between neurons. This change in weight values (new weight value is equal to old weight value + differentiation of error with respect to weights) is done starting from the output layer and proceeding toward the input layer. This process of learning new weights in this manner is termed as back-propagation. The process of deriving a new output, comparing with expected output and back-propagation, is repeated until the error is within acceptable limits (Figs. 1 and 2).

Artificial neural networks demonstrated a good capability to recognize patterns that were too complex to be symbolically defined as a knowledge piece. For example, there was initial success in recognizing hand-written digits. But researchers in artificial neural networks were not able to build solutions when confronted with problems that required more differentiation between attributes of the problem. Put in other words, artificial neural networks were limited to only a few layers of artificial neural nodes. The reason for this, especially in the supervised learning, lies with the back-propagation process. As the layers increase, differentiation of error with respect to weight results in a very small value. As such, the slope or level of change in weight value is too small to bring in a significant change in performance over the next iteration(s). This problem has been highlighted more in the current versions of artificial

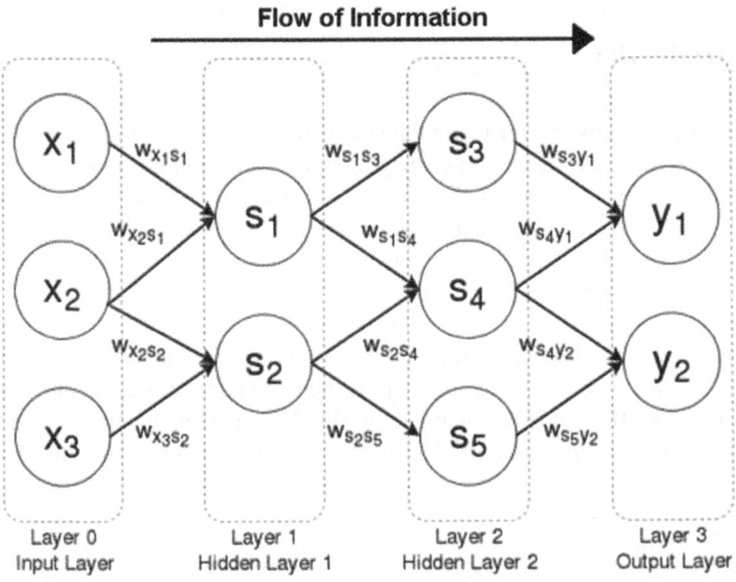

Fig. 1 Artificial neural network [3]

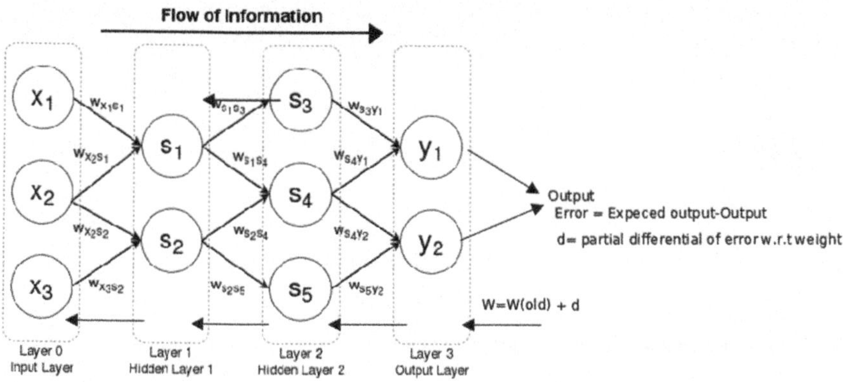

Fig. 2 Back-propagation [3]

neural networks termed "deep learning." In deep learning, more layers of artificial neural network neurons are made possible thus resulting in the nomenclature of "deep." The vanishing gradient problem in artificial neural networks was handled by an innovative use of the "expectation maximization" concept. Alex Krizhevskey and Geoffrey Hinton designed a convolutional neural network called "AlexNet," where in one iteration of the neural network they kept the weights fixed, while in the subsequent iteration, the error values were kept fixed. By this technique, the vanishing gradient problem was solved and multiple layers of neural network nodes were facilitated. Their approach won the ImageNet (ILSVRC) challenge of 2012 [6].

The attributes of a deep learning neural network are:

- Types of computational layers (of neural nodes): Convolutional, Auto-Encoder, Recurrent, LSTM, Belief, Adversarial, Dense
- Functions:
  - Activation: Mean Square Error, L2 loss, RELU, SIGMOID, TANH
  - Loss: Hinge loss, Cross-entropy
  - Optimization: ADAM.
- Normalization/Dropout layers: L1, L2.

Deep learning is advertised as doing away with "feature engineering" (other machine learning and artificial intelligence approaches require significant feature engineering). The statement claims that there is no need for a designer to analyze the domain and identify various classes of objects, attributes of each class and dependencies between the attributes as well as between different objects. In reality, the designer of a deep learning application still has to analyze the domain in detail. The activities in the two deep learning application developments described in this chapter will show what feature analysis is still required in the course of the development of deep learning applications.

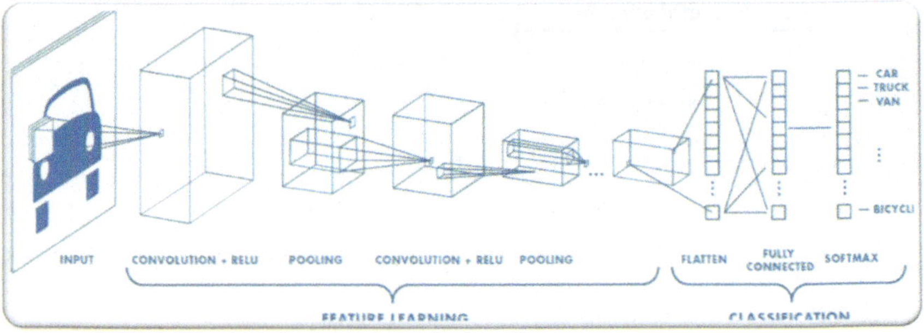

Fig. 3 Convolutional neural network [3]

The capabilities offered by the use of deep learning technologies have provided significantly better results in the fields of image classification/computer vision, natural language processing, speech recognition and predictive analytics. Real-time face recognition and handwriting recognition by a software have been made practically possible by the use of convolutional neural networks (Fig. 3).

A convolutional neural network (CNN) contains

1. convolution layers where the (mathematical) function defined in that layer works on the output of the previous layer's output
2. Max-pooling layers where only the maximum value of all outputs in the previous layer is allowed to pass through
3. dropout layers where some of the connections in the layer are probabilistically allowed to stop functioning and fully connected layers where all the outputs are considered.

Dropout layers are used to avoid overfitting of data, while fully connected layers allow classification among outputs. Max-pooling layers provide a reduction in the number of intermediate outputs being manipulated thus reducing the compute requirement. Surprisingly, it is also found that without Max-pooling layers, the CNN does not recognize images with the same capacity as with Max-pooling layers. This is one of the behaviors that need to be studied more because a Max-pooling layer actually reduces the details of the data being processed.

Every convolutional layer has a multitude of inputs taken in a matrix form. These are processed by using a filter of a particular size that is moved across the complete set of inputs. The filter focuses on a set of data points. The filter is moved across the layer of nodes with a defined stride. Each output from a particular application of the filter forms the input to the next layer.

AlexNet was constructed with five convolutional layers, max-pooling layers, dropout layers and three fully connected layers. This architecture was sufficient to classify one thousand categories. AlexNet used Restricted Linear Unit (ReLU) as the activation function, stochastic gradient descent as optimize function, used dropout layers and data augmentation techniques of image translations, horizontal reflections and patch extractions [6].

In the case of AlexNet and most other successful CNNs, the filter size is kept larger in the initial layers and becomes smaller in the layers closer to the output. This approach goes against the assumed process of identification where smaller components are identified first and used to build larger structures, finally resulting in an independent image identification. Additionally, use of the larger filters in initial layers hid that data/information from the subsequent layers. ZF Net tried to implement this idea of constructing images from smaller sub-components, by having the smallest filters in the first layer and increasing the filter size in subsequent layers. ZFNet was able to achieve identification results in lesser iterations than AlexNet [3].

1.1 Inception

InceptionNet [7] from Google set to rest the discussion on smaller versus larger filters by a new architecture where the inputs are provided to all the different layers, irrespective of their position. Similarly, outputs from all the layers are taken to a collator. An important point is that by this approach no data/information is lost to any level of analysis (all layers are open to all information). This is a non-sequential model and is called a graphical model.

Different filter sizes process the input and intermediate outputs. All the outputs are collected together before a final output layer. Many optimizations have been performed on inception architectures to reduce computational complexity: adding 1 × 1 convolutions, adding residual connections that add the output of the convolution operation of the inception module to the input. Inception approaches achieve a higher accuracy at an earlier training iteration (Fig. 4).

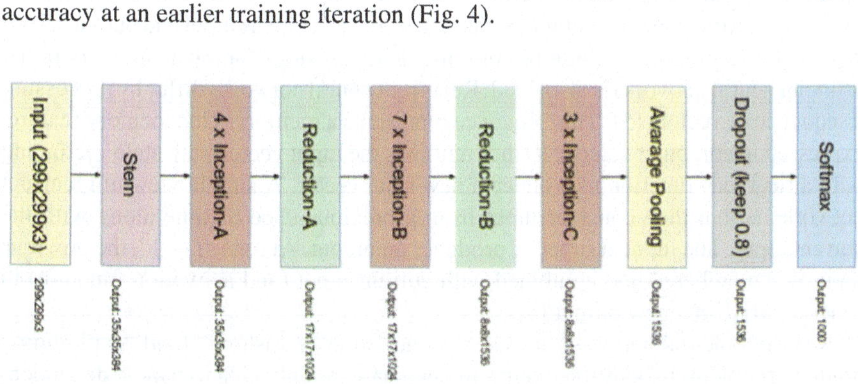

Fig. 4 InceptionNet [1]

1.2 Regional CNN

An approach that has had good success in identifying images along with their location is regional CNNs. While a CNN is primarily used for image classification, regional CNN (R-CNN) are better at object detection. In object detection, the location of the identified object is also detected; a set of bounding boxes is returned along with the object class that is identified. R-CNN uses a hierarchical approach to segregate overall image into parts [8]. This approach appreciates the fact that multiple scales in segmentation are required along with the acceptance that individual features cannot always decide the segmentation to be used. They also noted that there are many instances where there is a need to recognize the overall object before the individual objects can be grouped together [8]. Along with handling these matters, they have also considered the computational complexity of collectively processing a large number of pixels in order to ensure that any potential location of the object is not missed. The approach uses a data-driven segmentation approach and provides for increased diversity by using a set of complementary criteria to group as well as complementary color spaces with different invariance properties.

RCNNs are constructed as a combination of a proposal network along with a classifier network.

The progress in CNNs has led to a good adoption of this technique in real-time object classification and location identification.

1.3 Recurrent Neural Networks

Recurrent neural networks have been used to develop natural language processing models and speech recognition systems allowing a user to converse with the software in English or other natural languages. One of the advanced types of recurrent networks is long short-term memory networks, which have been successfully used to develop applications that can predict new values in a time series of observations. CNNs, like most other artificial neural networks, take a fixed size vector as input and output a fixed size vector as output. The number of computational steps (number of layers in the artificial neural network) is also fixed. Recurrent neural networks differ by processing a sequence of vectors [5]. This sequence represents a memory. This memory feature, expressed in computer science terms combines the input vector with state vector and with a (learned) function to produce a new state vector. A simple view of recurrent networks is that they consider input from a previous slice of time along with the current input. The input at time "t" produces an output. At time "$t + 1$", the previous input is remembered and combined with current input (and activation function) in order to arrive at a new output [5].

Recurrent neural networks are more prone to the "vanishing" (and "exploding") gradient problem. In addition, recurrent networks are only able to handle dependencies between data that are close together; they are not able to identify relationships between data that are significantly separated. Long short-term memory (LSTM) neural networks are a type of recurrent neural networks where these problems are handled. Memory is maintained and provided to all neuron cells. Each cell can now decide to add, forget or remember what is in memory. LSTM uses “gates,” where feedback of previous input is controlled. Only when it is relevant, is the feedback from a previous input allowed to affect the current computation. More specifically, there are gates which are composed of a sigmoid neural net layer and a pointwise multiplication operator. The sigmoid value will determine whether the input is to be let through for this neural node or not. Primarily, a forget gate layer that decides to omit or keep the previous input is used. For example, if a new instance of an entity is started to be processed, then the previous entities data are to be forgotten. An important variant of this forget gate is gated recurrent unit where forget and input gates are combined into one. This keeps a dependency between what is forgotten and what is input. This simpler version of the LSTM has a good performance and is widely used [2]. Other versions of LSTM include attention LSTMS and grid LSTMS [2] (Fig. 5).

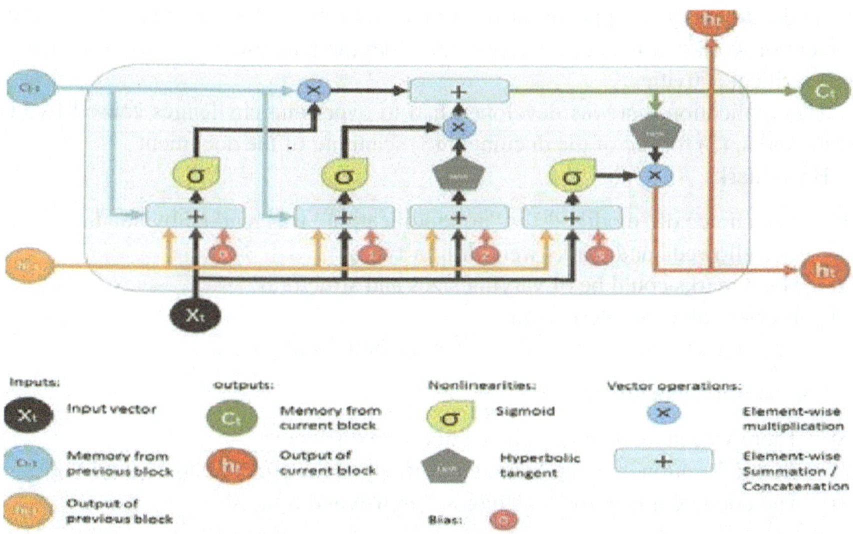

Fig. 5 Long short-term networks [2]

2 Application One: Image Recognition in a Document by Using Convolutional Neural Networks

The first deep learning task that is discussed in this chapter involves image recognition. The input was a document with multiple pages and containing multiple questions that applicants mark their choices on. The task was to identify choices by the applicants for each of these questions and to then do an analysis of the total result. The documents are given regularly to members of a particular governmental program. The questions are to check whether a particular attribute of the applicant has changed or not. Most of these attributes change only occasionally. So most (over seventy percent) of the filled in documents would finally be marked as not having any modification for all the multiple questions. All such documents need no further processing. Therefore, with the objective of effectively utilizing the manpower in the department conducting this program, the requirement was to develop a software application that could take the scanned images of the pages of each document and identify whether all the questions have been marked as "without modification" along with checking for the mandatory signature of the applicant at a specific location in the document. This application would thereby relieve the department of seventy percent of work that is not productive, allowing the manpower to be re-assigned to more critical activities.

The application that was developed had to overcome challenges caused by (1) hand marks, (2) format of the document (3) scanning of the document

Hand marks

(1) The choice of "modified" or "no modification" was marked by hand.
(2) The allowed hand marks were of four types.
(3) Hand marks could be of varying sizes and structures.
(4) Entries can be scratched out.
(5) The signature location is allowed at the bottom of a page.

Format of the document

(6) There was more than one choice question on a page.
(7) At one location, a barcode image overlapped the area of choice marking,
(8) The document is in two languages: English and Spanish.

Scanning issues (Pose problem)

(9) The page would not always be placed in the same location in the scanner
(10) The page could also be placed in a tilted manner.

2.1 Analyzing the Problem

Approach I—classical (symbolic) machine learning: A classical machine learning approach where the features are identified and learnt through inductive generalization of the results of a digital image pre-processing algorithm was considered. The complexity of the problem was a factor of the multiple key strokes that were possible for each hand mark, along with the possible x, y locations of the hand mark. This feature and image identification would then be contrasted with a free white space identification for the corresponding alternate choice to determine which of the two options had been chosen. Identifying the key strokes for each hand mark as a possible feature set was a complex task. The curves of a hand mark could differ from person to person. In addition, the location of these key strokes could differ even though it was bounded to be around the choice boxes. Third, the possible skewness of the scanned image led to an extrapolation of the locations and angles of these key strokes for each of the hand marks. Considering these three compounding factors, it was decided to use a deep learning approach to solve the task.

Approach II—transfer learning: One of the practiced approaches in industry to reduce the turn around time is to use transfer learning. Transfer learning is the process where a suitable model that has identified a super set/sub set of the images to be recognized is identified and loaded into the application, followed by a relatively short training with specific images that will enable recognition of the images relevant to the problem at hand. Along with training of the model, the deep learning architecture of the training application is modified as required. More particularly, when there is a good intersection of the trained images and the images to be recognized, then a couple of deep learning layers are added to the training architecture. These additional layers are normally added just prior to the classification (dense) layer.

Transfer learning was ruled out as an option, in this case as there were no available trained models that can recognize the hand marks that were used for this document.

Approach III—Training deep learning architecture to recognize the images: The option here is to create data sets of the four different hand marks and see whether the application can then recognize the presence of each hand mark at the choice of modified or of no modification for each of the questions. Each hand mark would be created in the various manners that they can be drawn. For example, one type of hand mark could be small, long, curved at the end, curved at the beginning, could have an acute or an obtuse slope. Similarly, another type could be perfect, crooked, elliptical, small, big. A third type was more difficult as they could be elongated in any orientation and be of any thickness. The fourth type could also be of different dimensions.

Another problem arose as part of this approach. There are multiple option boxes in a page. If the training is to identify a modified box chosen or a no modification box chosen, then the recognition stops in these pages with the identification of the choice for the first option box itself. The recognition does not proceed to check the other option boxes in the page. In other words, the deep learning application is trained to recognize a single option box for modified or for no modification. When the learned model is used for recognition, then the recognition stops with the very first option box with the result that it is a modified option chosen or a no modification option chosen. But what is required is for the deep learning application to recognize the choice for all the choice options in a page. To solve this, regional CNN was attempted. This required that the deep learning application has to be trained on the actual question for each option box. As this would require further training and subtle disambiguation, this approach was dropped. More specifically, each choice box had to be differentiated based on particular headings/words for each box. The issue was that these words could occur in other sentences too but not as a heading. This created a confusion for the application that was difficult to solve.

Approach IV—The next option taken was to consider the page as whole. For this, all combinations of choices for each option box were considered. One combination was all modified in the five option boxes of the page, while a second would be all no modification, followed by a single no modification, two no modifications, three no modifications and four no modifications. The single no modification had five possibilities (one for each option box) while the two no modifications had ten possibilities. Similarly, for the three no modifications and the four no modifications. A similar set of possibilities was done for the modified option. This was repeated for the other pages which had multiple options. Now the variations due to the different hand marks had to be considered for each of the four types. For each of the hand marks for each of the combinations possible, there had to be twenty samples created. In total, there were to be two thousand two hundred and forty samples to be created. Given the time constraint, the effort required to build the samples and the possible error that can happen while the samples are created, it was decided to abandon this approach.

Approach IV—The solution that was adopted was to decompose the problem (use "divide and conquer"). Instead of considering the complete page, the approach used was to find the independent entities of the task and approach them one at a time.

OpenCV tools were used to segregate the page into sections. Each section would pertain to one of the choice boxes. To identify each section, OpenCV's capability to identify straight lines was used. As each box was in a rectangle, OpenCV was used to identify these rectangles and the associated straight line. The straight-line location was used to cut the page into different sections. This was done for both training phase as well as for the recognition phase. So during training, samples were created for each section for each hand mark option.

To identify a section, the straight line that runs across the page, below a section was used. In addition, the identification of the choice box within a section helped in segregating the various sections from one another. By the use of OpenCV's functions, the page was split into sections where each section had exactly one choice box. Now, the overall task was broken down from identifying the choice in each of five choice boxes to one of identifying the choice in a single choice box and then accumulating results across the sections.

2.2 Handling the Challenge of Skewness

The second challenge was in handling skewness of the scanned image. When the page was placed in a scanner to generate an image, it could be placed in a misaligned manner. This misalignment caused problems in recognition of images within the document. This problem is classified as the pose (translational invariance) problem in the literature of deep learning previous research approaches this problem using Hough transform. The pose problem is a current research topic [4]. Human eyes seem to easily adjust the perception to accommodate tilts in the object being recognized. This accommodation is difficult to do for computer vision tasks. In this application, a combination of OpenCV's line finding functions along with a calculation of the angle of the lines to determine whether the page is tilted or not is used. First, a straight line at the top of the page is identified. This is used as a reference. Next, straight lines at different sections of the page are identified and analyzed to determine the angle of the line. The angle was then corrected to be parallel to the top straight line of the page. This straightening process was repeated at every section of the page as the angle of skewness was found to vary across the page. In this manner, the scanned image was straightened. This skewness correction was done as a pre-processing prior to training/recognition.

2.3 *Architecture of CNN*

Layer (type)	Output shape	Param #
conv1d_5 (Conv1D)	(None, 12, 64)	29,312
max_pooling1d_5 (MaxPooling1)	(None, 12, 64)	0
dropout_10 (Dropout)	(None, 12, 64)	0
conv1d_6 (Conv1D)	(None, 10, 64)	12,352
max_pooling1d_6 (MaxPooling1)	(None, 10, 64)	0
dropout_11 (Dropout)	(None, 10, 64)	0
conv1d_7 (Conv1D)	(None, 8, 64)	12,352
max_pooling1d_7 (MaxPooling1)	(None, 8, 64)	0
dropout_12 (Dropout)	(None, 8, 64)	0
conv1d_8 (Conv1D)	(None, 8, 64)	4160
max_pooling1d_8 (MaxPooling1)	(None, 8, 64)	0
dropout_13 (Dropout)	(None, 8, 64)	0
conv1d_9 (Conv1D)	(None, 8, 64)	4160
max_pooling1d_9 (MaxPooling1)	(None, 8, 64)	0
dropout_14 (Dropout)	(None, 8, 64)	0
flatten_2 (Flatten)	(None, 512)	0
dropout_15 (Dropout)	(None, 512)	0
dense_5 (Dense)	(None, 10)	5130
dense_6 (Dense)	(None, 1)	11

Total parameters: 67,477
Trainable parameters: 67,477

Non-trainable parameters: 0.

Five convolutional layers were used to identify the key features of the image. Each convolutional layer was followed by a Max pooling layer and a dropout layer. Finally, a fully connected layer was used for classification of the image.

The input layer took the image as a set of 64 × 64 pixels. The filter sizes used were

- Layer 1: 5 × 5
- Layer 2: 5 × 5
- Layer 3: 3 × 3
- Layer 4: 3 × 3
- Layer 5: 1 × 1.

Stride was maintained as 1 × 1 for all layers.

Learning rate was maintained at 0.1.

The MaxPooling layers (after each convolutional layer) had a size of 64 × 64 pixels.

A dropout layer of was also introduced after each MaxPooling layer. At the output, two dense layers were used in order to produce the best value as the classification of the image.

In the course of the project, different filter sizes and different number of layers were tried. The other architectures experimented with were:

Layers:

- Six layers of convolutional layers with corresponding MaxPooling and dropout layers
- Four layers of convolutional layers with corresponding MaxPooling and dropout layers
- Seven layers of convolutional layers with corresponding MaxPooling and dropout layers.

Filter Sizes:

- Seven by seven followed by five by five followed by three by three
- One by one followed by three by three followed by five by five (similar to ZFNET).

Different loss functions were also experimented with:

- Linear
- Sigmoid
- RELU.

Training:

Data set creation: Due to the multiple locations where a hand mark can be made within the option box, the training set was created with a distinct space between each mark: top left, top center, top right, mid left, mid center, mid right, down left, down center, down right, along with random marks between these nine locations. This process was repeated for all the option boxes. Then, these were scanned and converted into images. Each set of twenty images was stored into appropriately labelled folders. Twenty samples were created for each category.

The different architectures were run on separate computer systems, in order to more quickly evaluate the results.

Variations in the loss functions were done in a second iteration.

Different filter sixes were tried in a third iteration.

A mix of architecture and loss function was evaluated in a final iteration

- ($5 \times 5, 5 \times 5, 3 \times 3, 3 \times 3, 1 \times 1$ filter size) with RELU versus same architecture with Sigmoid.

2.4 *Prediction/Testing*

The first level of predictions was done with a system flow that did not include page skewness alteration. This led to good results initially but were followed by bad results in later trials. Then, after the problem of page skewness was identified and solved, the results were more consistent. For prediction, a different set of people were requested to make the hand marks at random locations.

2.5 *Results*

Our application passed all tests with a 99.9% accuracy. Tests were conducted with different types of hand marks, done by different people. Mixed mode of hand marks were also tested, where some choice boxes had (for example) check marks and others had circle hand marks. The only test case where the results were not always correct was in the location where a barcode image covered part of the choice box and the hand mark used was overlapping this area of the choice box.

The application was demonstrated using scanned images of documents that had external people recording their choices for each section of the documents. It provided satisfactory results for the end user.

3 Application 2: Predicting Equated Monthly Instalment Payments

The second deep learning application that is described in this chapter was developed for a challenge faced in financial institutions. The financial institution lends money to their customers to allow them to procure vehicles. The customer pays back the loan amount in Equated Monthly Instalments (EMI). Unfortunately, customers default on their payments. If the defaulting payments continue over specified time periods, then the account is termed as a non-performing asset (NPA). All financial institutions would like to reduce their NPA. There is therefore a need to predict and take appropriate action on possible NPAs. The requirement given was to predict with maximum accuracy, the amount of payment that every one of their customers will make in an upcoming month.

Initial discussions with the financial institution highlighted that the payments were related to income, business expenses, family expenses, other investments done by the end user, as well as business trends, changes in government policies, natural calamities, economical and financial changes in the world. The organization did not contain the details of their client's demographics, family details, illness history, educational expenses, other investments, nor was there a means to determine the business that the client was engaged in. The primary data that was made available was the customer's history of EMI payments. The task was then to predict the subsequent month's EMI payment of a customer, based on the history of payments made by that customer. Another attribute that was seen to have an impact on payments made was the type of vehicle that the customer had purchased. This data was also provided to be used to determine the next month's EMI.

There were over seventeen thousand customers. The data provided were payment data over thirty months, for seventeen thousand customers. Payment data were broken down into total loan amount, number of months of loan, amount due per month and amount actually paid. The amount due for every month varied based on the type of loan that the customer had taken. Therefore, this became a specific attribute for every instance.

3.1 Analyzing the Problem

The problem was identified as a "multi-variate, multi-time" series problem. One customer's monthly payment sequence was one time series. The payments would depend on previous payments. The capability to make a payment of a particular amount would decide the capability to make a similar payment in the next month. There could also be cases where the customer would opt to make a larger payment in one month and pay less or nil in a subsequent month. Similarly, different customer payments would result in different patterns of time series. The task was therefore not a simple time series but a multiple time series. The architecture best suited to model memory-based behavior is a long short-term memory (LSTM). In the area of LSTM multi-time series is an ongoing research topic. Currently, the approach is to add an identifier for each time series allowing the deep learning system to learn and then predict each time series accordingly.

3.2 Data Representation

In the data that were shared, every customer had thirty-two attributes relating to location, customer, marketing personnel, status of account, amount receivable, arrears and payments dues over months.

Of these fields, the ones that were relevant for prediction were selected as:

Loan_Amount, Required_monthly_amount, arrear, description of item, due_amounts (over months) paid_amounts (over months).

Of these, one customer could have data as 450,000, 21,000 0, A—Used, 21,000 21,000 21,000 21,000 21,000 21,000
Another customer could have data as 700,000, 27,100, 0, E—Used, 27,100 27,100 27,100 27,100
A third customer could have 913,040, 31,390, 4517, E—Used, 31,390 62,900 31,390 32,000 31,390 0 31,390 32,000.

(Above data are changed in order to maintain the privacy of the organization but reflect the actual data).

Only three data have been shown. Considering the variation in all these data, over seventeen thousand customers created a challenge. The first approach to normalize the data was to replace the numbers by percentages. So, the percentage of payment made per month of the due amount was used. (The due amount was used for the calculation but was not stored in this new representation.) This converted the data to a format that could be analyzed better. The result of this conversion created data that looked like (for the above three cases)

450,000, 21,000 0, A—Used, 100 100 100
700,000, 27,100, 0, E—Used, 100 100
913,040, 31,390, 4517, E—Used, 200, 100, 0, 100.

This converted varied numbers of 21,000, 27,100 and 31,390 to a single percentage value of 100. A similar result was seen for other customer records. A few examples are given below

Vehicle id	% in M1	% in M2	% in M3	% in M4	% in M5	% in M6	% in M7	% in M8	% in M9	% in M10	% in M11
222	100	100	100	100	100	100	100	100	100	100	2920
235	200	0	100	210	0	110	100	0	100	110	100
547	100	100	0	0	0	0	300	0	2620	0	0
547	100	100	0	0	0	0	300	0	2620	0	0
547	100	100	0	0	0	0	310	0	2620	0	0
547	100	100	0	0	0	0	310	0	2620	0	0

The first two rows show difference in payment details, while the next two are the same "payment pattern" and the last two form another payment pattern. Thus, converting the payment to a percentage form resulted in the similarities between payments becoming explicit. This was a major step toward identifying patterns and making predictions. The percentage calculation and subsequent pattern identification formed a major part of the "data pre-processing" in the developed application.

The next challenge that was overcome referred to the various types of attributes. One set of attributes was the payment in every month along with the due amount every month, while another set of attributes concerned the initial loan sanction details, and a third set referred the type of and model of vehicle that was purchased with the sanctioned loan. All these attributes had to be converted to a common range and unit in order to ensure that the specific values did not bias the learning and subsequent prediction. It was also noted that the requirement was to predict the classification of the customer's payment into one of the possible percentage values. Using the fact that there was to be a categorical classification, one-hot encoding was used to represent the attributes. By using one-hot encoding, the need to do categorical classification and the need to represent various types of attributes were met. While the solution was convenient from the perspective of a common range of values for all the attributes, there were other challenges caused by this approach. For one, the memory requirement increased significantly, as each of the attribute values were converted to a one-hot encoding. This meant that training with all the payment patterns for the seventeen thousand customers could not be done in one instance. The payment patterns were grouped and trained in batches. Each of the training would then create a separate model. This meant that during prediction also the payment pattern of the customer has to be first identified and then the appropriate model for this pattern would be used to make the prediction. One-hot encoding also created a problem of mandating a particular number of columns for every data set that was used for making a prediction. If the number of columns was not consistent with the columns used for the training time, then the application would throw an error. So, during prediction of payment, the input data had to be analyzed to ascertain the number of columns and then dummy data had to be added if the columns did not match. One-hot encoding was used for this task with these additional support modules implemented.

3.3 *Design of the LSTM*

The LSTM software development to predict the Next month EMI payment by customers was divided into four primary stages:

(a) Feature exploration: Define Relevant features
(b) Prototyping and training: Train a model prototype
(c) Model selection and validation: Perform model selection and fine tuning
(d) Productization: Take the selected model prototype to production.

Feature exploration involved four tasks of

(i) Data analysis: Analyzing format of data and related values provided by the financial institution
(ii) Development of a program to pre-process the data: Organizing data, converting data and re-arranging data. In this, the payment amounts were made a percentage of the due amount for each month
(iii) Program to identify data features: Categories of payment patterns were identified. First, it was decided to use a variation of ten percent in payment as an indicator of a different payment pattern. More specifically, if one customer had made a payment of 100% in a particular month, another customer who made a payment of less than 91% would be considered to be a different payment category. So, finding payment patterns meant that the payments made in every month by one customer were compared with the payments made by other customers. If in any month, there was a variation of 10 or more percentage, then this customer's payments were considered to be a new payment pattern. Before arriving at "ten percentage" as the proper range to use, payments were analyzed with the following variations:

    - All thirty months as a payment pattern
    - Twelve months as a payment pattern
    - Three months as a payment pattern.

    The percentage values were varied across:

    - Ten percent
    - Five percent
    - Two percent.

    All these payment patterns were given for training to the LSTM architecture. The best and most consistent results were got for the twelve months ten percent combination. In this manner, thirteen thousand three hundred and forty-two payment patterns were identified over the thirty months of payments, of seventeen thousand two hundred and seventeen customers.
(iv) Identify and predict patterns: Build a (deep learning) software that can identify the correlations between the categories.

It was observed that customer payment patterns were changing over time. Keeping the impact of this on the required prediction, it was decided to focus on the customer behavior at the end of the time period of the data that were shared by the financial institution. Here, the number of customer records that had twelve or more months of data was eight thousand five hundred and sixty-eight.

The LSTM architecture to learn and then predict customer payments had six thousand one hundred and eighty-one parameters. There were four LSTM layers, each followed by a dropout layer. Finally, two dense layers were used for the classification.

Layer (type)	Output Shape	Param #
LSTM_1 (LSTM)	(64, 10, 8)	4448
DROPOUT_1 (Dropout)	(64, 10, 8)	0
LSTM_2 (LSTM)	(64, 10, 8)	544
DROPOUT_2 (Dropout)	(64, 10, 8)	0
LSTM_3 (LSTM)	(64, 10, 8)	544
DROPOUT_3 (Dropout)	(64, 10, 8)	0
LSTM_4 (LSTM)	(64, 10, 8)	544
DROPOUT_4 (Dropout)	(64, 8)	0
DENSE_1 (Dense)	(64, 10)	90
DENSE_2 (Dense)	(64, 1)	11

Total parameters: 6181
Trainable parameters: 6181
Non-trainable parameters: 0

Recognizing multi-time series: As there were many customers making payment, there were different patterns of payment. This meant that there were different time series that had to be learnt by the LSTM. This is one of the research challenges of LSTMs; how one architecture can learn different series. Based on the advice of research articles on multi-time series, an identifier was introduced for each time series. The pre-processing algorithm identified different time series based on its payment pattern ID. This payment pattern ID was added to the customer's payment history as a time series identifier. As stated before, there were five thousand and twenty-four patterns discovered through the data pre-processing. Training was done with the time series identifier and all the "percentage converted" customer payment data. Surprisingly, the results were not satisfactory. Around fifty percent validation was only achieved. Variations in the data format were now explored. Vehicle ID was removed, customer data for shorter ranges were tried, dropout was decreased, and payment pattern ID was removed. The best results were arrived at when the time series identifier was removed. A 99% accuracy was reached with this architecture and framework. An intuitive conclusion is that the statistically arrived at payment pattern classification was a misleading indicator. It biased the LSTM in the wrong manner. Removing this pattern identifier seems to have allowed the LSTM architecture to arrive at its own identifier between the five thousand and twenty-four different payment patterns. This is a point to be probably studied further.

3.4 *Results*

A 99% accuracy was achieved in predicting the next month's EMI payment for all customers, who had a history of payment over ten months. Prediction was done for all customers and validated using the following month's actual payment. There were cases where the payment pattern changed in the following month. Leaving aside these cases the application was able to predict with 99% accuracy the next month's payment (within a range of ten percent). The multi-variate multi-time series was handled successfully by the LSTM that was developed.

4 Review of the Two Applications

(1) The approach that worked for identifying the hand marked choice exhibited the good usage of divide and conquer. This has to be done with care to ensure that only independent components are addressed separately, without ignoring any underlying dependencies that may be present.
(2) The LSTM project improved its results to a satisfactory level after the time series identifier was removed from the data set. This seems to imply that the payment pattern ID that was statistically calculated and added to the data set was a wrong bias, encouraging the application to predict wrongly. Therefore, the topic of identifier for a multi-time series LSTM prediction needs to be researched better.

Developing a deep learning application involves a high amount of engineering:

- Analyze the problem: As the problem is one that needs intelligence to solve, the analysis phase is significantly more demanding than other software tasks.
- Possible solutions: A set of possible solutions emerge after the analysis with no clear distinction between them. All have to be attempted and improved upon.
- Experimental results: Results at each stage have to be recorded granularly and analyzed to decide the architecture/design for the next iteration toward deriving the required result.
- Improve architecture: Analyzing performance is used to decide how the architecture can be changed. Certain combinations may not be possible/not recommended, while others have to be tried and checked.
- Perseverance: Different data pre-processing, designs, parameter values, attribute changes and more, finally, give a good result.

References

1. Raj, B. (2018, May). A simple guide to the versions of the inception network. Medium. https://towardsdatascience.com/a-simple-guide-to-the-versions-of-the-inception-network-7fc52b863202.
2. Colah blog. (2019, December). Understanding LSTM Networks. http://colah.github.io/posts/2015-08-understanding-LSTMs/.
3. Deshpande Adit. The-9-Deep-Learning-Papers-You-Need-To-Know-About https://adeshpande3.github.io/The-9-Deep-Learning-Papers-You-Need-To-Know-About.html.

4. Elton, D. (2017). http://moreisdifferent.com/2017/09/hinton-whats-wrong-with-CNNs.
5. Andrej Karpathy blog. (2015, May). The unreasonable effectiveness of recurrent neural networks. http://karpathy.github.io/2015/05/21/rnn-effectiveness/.
6. Krizhevsky, A., Sutskever, I., & Geoffrey, H. (2017). Imagenet classification with deep convolutional neural networks. *Communications of the ACM, 60*(6), 84–90. https://doi.org/10.1145/3065386. ISSN 0001-0782.
7. Szegedy, C., et al. (2015). *Going deeper with convolutions*. CVP Computer Vision Foundation.
8. Uijilings, J. R. R., et al. (2012). *Selective search for object recognition*. Technical report, 2012 IJCV.

Speed, Cloth and Pose Invariant Gait Recognition-Based Person Identification

Vijay Bhaskar Semwal, Arghya Mazumdar, Ashish Jha, Neha Gaud, and Vishwanath Bijalwan

Abstract Gait is very important to identify person from distance. It requires very less interaction with human participants. Gait is considered the popularly known visual identification technique. The major challenges associated with gait-based person identification are high variability, gait occlusion, pose and speed variance and uniform gait cycle detection, etc. In this research work, the CASIA-A, B and C data set is explored for the view, cloth and speed invariant person identification to address the challenged associated with gait-based person identification. In this work, the very important technique of computer vision for object identification is being explored. It included feature extraction techniques, namely gait energy image(GEI) for cloth invariance, histogram of gradients(HOG) for multiview invariance and Zernike moment with random transform for crossview invariance. To classify data, SVM, ANN and XGBoost-based machine learning algorithms are used on the CASIA gait data set and achieved 99, 96 and 67% identification accuracy, respectively, for three different scenarios of invariance, i.e. speed, cloth and pose.

Keywords Human locomotion · View invariance · Speed invariance · Gait identification · Machine learning

---

This paper is the outcome of work carried out under the project tilted **Development of Computational model for bipedal walking trajectories generation and analysis of different gait abnormality**. The project is supported by SERB, DST (Government of India) under Early Career Award to Dr. Vijay Bhaskar Semwal, PI of project with DST No: ECR/2018/000203 ECR dated 04-June-2019.

V. B. Semwal (✉)
MANIT Bhopal,Bhopal, India
e-mail: vsemwal@manit.ac.in

A. Mazumdar · A. Jha
NIT Rourkela, Rourkela, Odisha, India

N. Gaud
SCSIT DAVV Indore, Indore, India

V. Bijalwan
Institute of Technology Gopeshwar, Gopeshwar, India

M. Pandey and S. S. Rautaray (eds.), *Machine Learning: Theoretical Foundations and Practical Applications*, Studies in Big Data 87,
https://doi.org/10.1007/978-981-33-6518-6_3

1 Introduction

Gait is the process in which upper and lower body act in unison. It can be loosely understood as person's way of walking. The gait cycle has two distinct phases. One is stance, and other is swing. The entire gait cycle can be divided into eight subphases as shown in Fig. 1. In stance phase, the gait cycle has initial contact, loading response, mid-stance and terminal stance. In swing phase, the gait cycle has preswing, initial swing, mid-swing and terminal swing [1]. In the gait cycle, hip, knee and ankle moves are distinct way to produce gait. We aim to differentiate and identify people based on their gait [2].

Human gait is used to identify the person from distance, and it is unobstructed biometric trait process for person identification. Gait analysis is very important in surveillance, identification purpose and security infrastructure system [3]. Right now we have fingerprint, face recognition for biometric recognition but none of the techniques works when the subject to be identified is at a distance. Gait is the only biometric trait that can identify subject at a distance [4]. Gait analysis is done for medical purposes too where it can be used for early detection of gait abnormalities including Parkinson disease. The gait study further can be utilized for generation of robot walking trajectories [5]. Gait is also used for planning the path of humanoid robot [6].

Gait is considered as behavioural biometrics which has the highest collectabilty as shown in Tables 1 and 2. Gait suffers with low permanence at early state of learning. A learner's gait dynamics can change drastically within a short period of time as

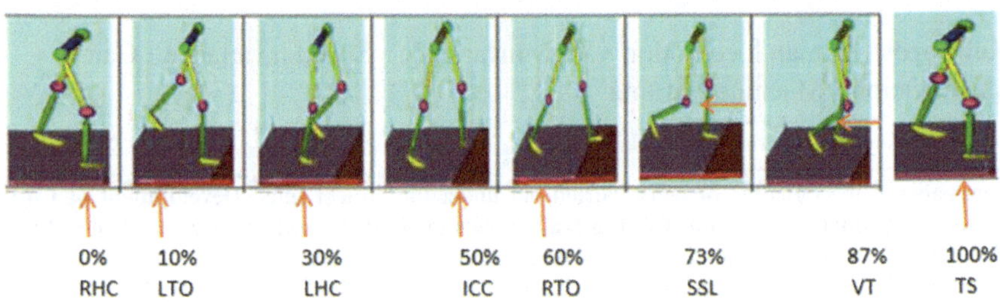

STANCE(60%)				SWING (40%)		
Loading Response [0-10] %	Mid Stance [10-30%]	Terminal Stance [30-50%]	Pre Swing [50-60%]	Initial Swing [60-73%]	Mid Swing [73-87%]	Terminal Swing [87-100%]
Double Support	Right Single Support		Double Support	Left Single Support		

RHC - Right heel contact , LTO - Left toe off, LHC- Left heel contact, ICC- Initial contact of the contralateral limb, RTO – Right toe off,SSL - Swimming limbs opposite to stance limb,VT- Vertical tibia, TS- Terminal swing.

Fig. 1 Different subphase of one complete human gait cycle

Table 1 Different physiological-based biometric characteristics

Biometric	Universality	Distinctiveness	Permanence	Collectabilty	Circumvention
Face	High	Medium	Medium	High	Low
Iris	High	High	High	Medium	Low
Palm print	High	High	Medium	Medium	Medium
Fingerprint	High	High	Medium	Medium	High
Retina	High	High	High	Low	Low

Table 2 Different behaviour-based biometric characteristics

Biometric	Universality	Distinctiveness	Permanence	Collectabilty	Circumvention
Gait	Low	Medium	Low	High	Low
Speech	High	Medium	High	Medium	Medium
Signature	High	Medium	Low	Medium	Low
Keystroke	High	Medium	Low	Medium	Low
Device Uses	Low	Medium	Low	High	Low

the learners gets accustomed to the environment being used. Once it is acquired and accustomed, it is very less likely to change with time [7]. Figure 1 shows the different subphases of one complete gait cycle. The one human gait cycle consists of two broad phases: one is single support phase (SSP) when one foot will be in air and another will be on ground, and another phase is double support phase (DSP): when both foot will be place on ground [8]. The DSP is observed very less during normal walk and quantitative; it is only 10–12% one gait cycle. One healthy person can complete the one gait cycle in between 0.52 and 1 s [9]. The one complete gait cycle cab be further divided into seven different linear subphases [10].

The rest of the paper is organized into five sections. Section 2 presents the literature review about the gait recognition work done so far. Section 3 is presenting our proposed view, cloth and speed invariant method. Section 4 is result and discussion section and presents the performance study of all the algorithms used for classification, and, finally, Sect. 5 presents the conclusion and future research work.

2 Literature Review of Related Work

In the past, various spatial and temporal features have been used for gait-based identification. A number of models have been proposed including vision, sound, pressure and accelerometry models [11, 12]. Gait signals can be complex which make gait-based identification tough. Two different approaches have been developed for image-based gait recognition. One is model-based method which computes the model features by fitting model to the image, and other is appearance-based strategies [13].

Many researchers have tried to improve the accuracy of machine vision-based gait identification [14]. Researchers have proposed a method in which the gait features are obtained by calculating the area of the head, arm swing and leg swing regions. This method works only in speed invariant human identification problem. Several researches have used speeded up robust features to depict the trajectories of the different parts of the body, but this method works only if there is only one moving object [15]. Hence, it is not suited for crowded scene. Kusakunniran [21] proposed a technique in which space-time interest points are obtained from a raw video sequence. The advantage of this method is that the time complexity occurred by pre-processing of the video is removed. Wang et al. [16] developed a temporal template called as chrono gait image. This method requires the less computational cost to preserve the gait cycle information.

The approach (Fig. 2) is based upon the notion that the boundary of the silhouettes (Fig. 3) will contain maximum information, and this would hold true for any kind of gait condition including the person in a different attire, with a bag or walking at

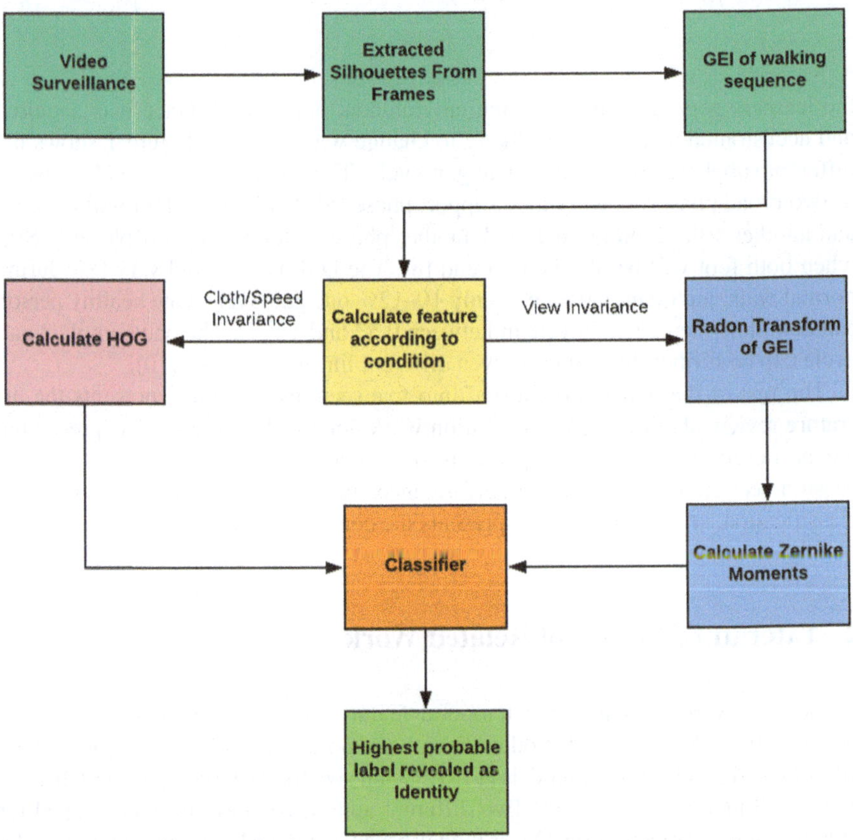

Fig. 2 Process flow diagram

Fig. 3 Gait silhouette

different speeds. However, since, the silhouettes represent a walking sequence over a particular set of frames, we need to convert them into a single frame by using GEI [17].

2.1 GEI

Human walking is a repetitive motion which remains same up to a frequency. The order of positions of the limbs is the same in each cycle that is the legs move forward and the arms backward. This order is not important in feature extraction for gait recognition and so can be combined to form a spatio-temporal representation of a single image by using GEI [18, 19]. The greater the pixel value in the GEI, more is the frequency of occurrence of the human body.

Pre-processing steps include extracting region of interest of the silhouette image and finding out the centre of the human [20]. Although it might seem implicit in nature, the region of interest found out here is based on a novel approach which is termed as pixel approximation method Algorithm 1 which is then used to calculate the GEI.

2.2 HOG

HOG descriptors can be used to extract boundary information in a swift manner from the GEI images (Fig. 4). It is a feature descriptor technique used in image processing, mainly for object detection. The distribution [histograms) of directions of gradients is used as features here. Gradients (x and y derivatives) of an image are pivotal because the magnitude of gradients is large around edges and corners (where intensity abruptly changes). Here, nine orientations are considered with pixels per cell being (16, 16) and cells per block is (2, 2), for each image. To calculate the final feature vector for the entire image, the individual vectors of each path are flattened

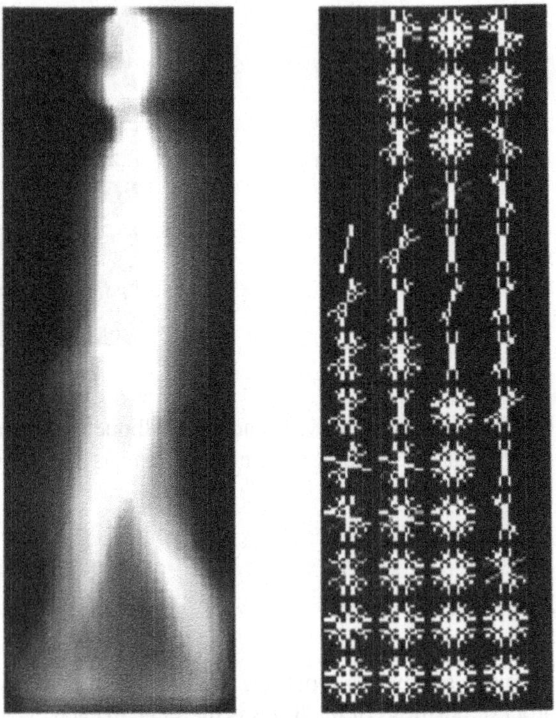

Fig. 4 GEI (Left) used in obtaining HOG features (Right)

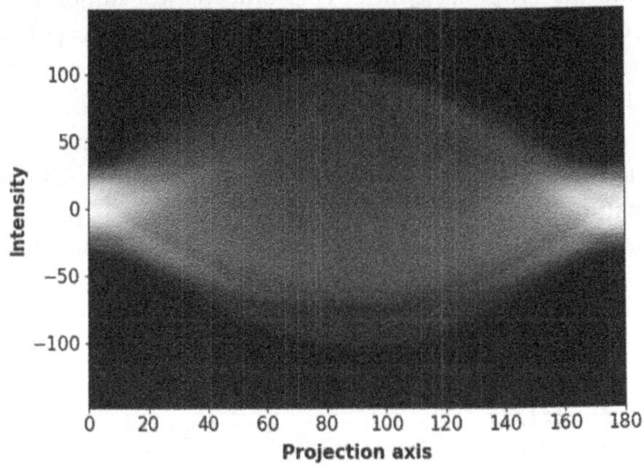

Fig. 5 Radon transformed GEI image

into a array having 1420 elements with the original image size being 210×70, thus representing a 90% reduction in size of feature vector. Benefits of selecting this feature are that it is compact, fast to calculate, scale and translation invariant and work appreciably for silhouette images.

Algorithm 1 ROI calcuation with centre of Image location for GEI building

Result: GEI is build on passing silhouette images of a walking sequence
Input: Image (i) with width (w) and height (h)
Output: GEI
Procedure

- **Set four pointers** $up \leftarrow 0, down \leftarrow 0, right \leftarrow 0, left \leftarrow 0$ **to capture the delimiters of the silhouette in four directions.**
- **Increment the pointers until a white pixel is found so that the rectangular boundary is obtained**
- **Crop the image now by making use of the pointers**

$pixelCount \leftarrow []$
index $\leftarrow$ **right -left**
pixelCount $\leftarrow$ $Non\ Zero\ Pixels\ In\ Cropped\ Image$ **countAll** $\leftarrow \sum$**(pixelCount)**
pixelPercent = count / countAll foreach count in pixelCount
countPercentSum $\leftarrow$ **0, minTheta** $\leftarrow$ **1,bestIndex** $\leftarrow$ **0**
index,val in pixelPercent:
tmp $\leftarrow$ **|0.5 - countPercentSum|**
if tmp <minTheta then
minTheta $\leftarrow$ **tmp**
bestIndex $\leftarrow$ **index**
end if
end for

- **Centered Human is Image with new dimensions bestIndex, w-bestIndex**
- **Calculate mean of Centered Human for a Time frame(T) using** $G(x, y)$**=1/N** $\sum_{t=1}^{t=N}$ **B(x,y,T)**

HOG can not only capture the outlines of the different clothes and bags in the upper torso , by combining it with the Shannon entropy (GEnI) which finds out the uncertainties of the pixel positions of the lower torso (legs), and the variations in the walking speed are also captured.

2.3 *Radon Transform and Zernike Moments*

Difficulty comes with the third aspect when the subject is viewed from different angles and trained on a subset of them (multiview) or viewed at a different angle from which they are trained on (crossview) since HOG descriptors although being scale and translational invariant are not rotational invariant. To overcome this problem, the help of radon transform is taken from the tomography field which takes the projection

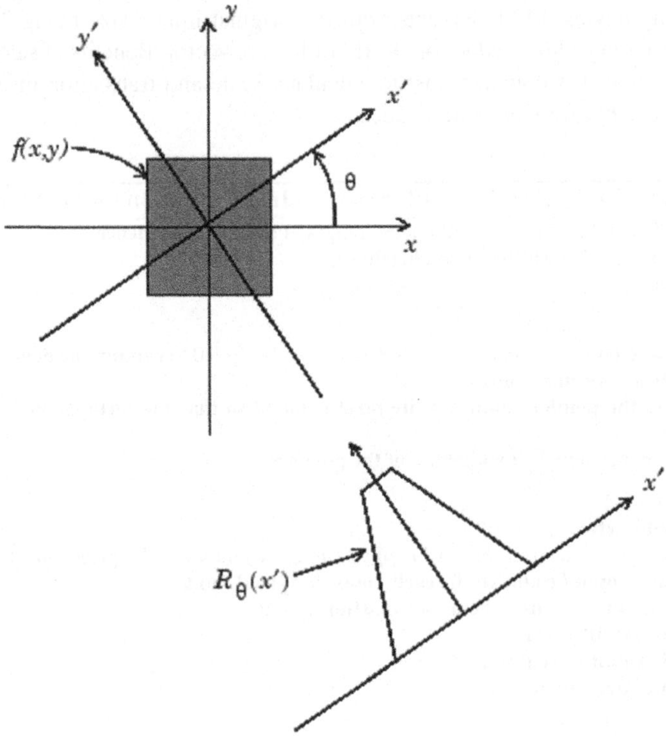

Fig. 6 Random transform

of a 2D image along a direction to give it a 1D profile something resembling taking an X-ray scan on the edge of the image.

A projection of a two-dimensional function $f(x, y)$ is a set of line integrals. The radon function computes the line integrals from multiple sources along parallel paths, or beams, in a certain direction. Projections can be computed along any angle theta θ. The sinogram (term used to describe the data after radon transform) has been computed here as θ ranging between 0 and 180 (Fig. 6) with the number of values generated equal to the image width. The formula of computation is as follows:

$$R_\theta(\acute{x}) = \int_{+\infty}^{-\infty} f(\acute{x}\cos\theta - \acute{y}\sin\theta, \acute{x}\sin\theta + \acute{y}\cos\theta)\mathrm{d}\acute{y} \quad (1)$$

where

$$\begin{bmatrix}\acute{x}\\ \acute{y}\end{bmatrix} = \begin{bmatrix}\cos\theta & \sin\theta\\ -\sin\theta & \cos\theta\end{bmatrix}\begin{bmatrix}x\\ y\end{bmatrix}$$

This RGEI(Radon GEI) is then used to obtain the Zernike moments which are rotational invariant (Fig. 5) unlike HOG but do not capture boundary information so well nor are substantial enough to be fed into a classifier. Complex Zernike moments are constructed using a set of complex polynomials which form a complete orthogonal basis set defined on the unit disc $(x^2 + y^2) \leq 1$. They are expressed asA_{pq} two-dimensional Zernike moment:

$$A_{\mathrm{mn}} = \frac{m+1}{\pi} \int_x \int_y f(x, y)[V_{\mathrm{mn}}(x, y)]^* \, \mathrm{d}x \, \mathrm{d}y \tag{2}$$

The moments were calculated here according to the maximum radius for the Zernike polynomials being 85 pixels and degree being 8 to get 27 values for each feature vector of the image. Advantages of using Zernike moments include simple rotation invariance, higher accuracy for detailed shapes, orthogonal which means it has less information redundancy and better at image reconstruction. It also gives rotational invariance unlike HOG.

2.4 *Classifier Model*

The final step is to feed the mentioned features into a classifier according to the condition for prediction of identity of individuals through their walking sequence obtained from the video frames. For investigation purposes, different classifiers have been trained ranging from traditional methods like artificial neural network (ANN) (Fig. 7) to support vector machine (SVM) to newly discovered techniques including XgBoost which has obtained higher accuracy in recent classification challenges. Detailed description of models is given in Table 3.

3 Result and Discussion

3.1 *About the Database*

Casia-B Database
CASIA gait database contains the human walking pattern of 124 subjects. It has been made open source by Centre for Biometric and Security Research, CASIA. For each subject, video is captured from 11 different angles. Also each subject is captured in three different variations, namely normal walking, walking with bag and walking with different clothing conditions.

Casia-C Database
CASIA-C contains 153 subject and considers four variations, namely normal walking, slow walking, fast walking and walking with a bag. It was captured via infrared

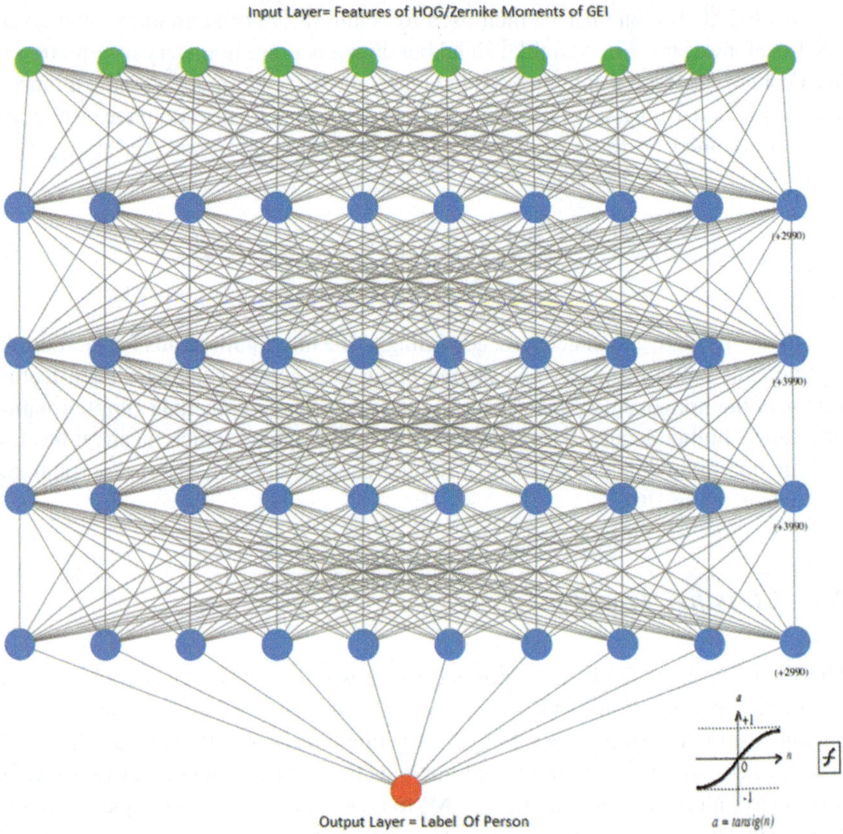

Fig. 7 Structure of neural network (plotted in graphviz) + denotes additional neurons not shown

Table 3 Description of model parameters

Model used	Parameters
SVM	Kernel = Linear, penalty parameter (C) = 1 height
	Solver = Limited memory Broyden–Fletcher–Goldfarb–Shanno (LBFGS),
ANN	Hidden layers=4,
	Neurons = (3000, 4000, 4000, 3000),
	Alpha (Regularization L2) = 0.001
XGBoost	Booster = Dart

camera and has been made open source by Centre for Biometric and Security Research, CASIA.

From a machine learning domain, these data sets represent a multiclass classification task with balanced classes since for each class there is only one GEI created from the frames. Thus, the results are given in macro average format which gives equal importance to all classes which is applicable here because it is balanced in nature.

3.2 *Speed Invariance*

Experimental Condition
CASIA-C data set was used for this instance, and the training set consisted of two normal walking sequences (fn), one fast walking sequence (fq) and one slow walking (fs) sequence. Validation set consisted of one normal walking sequence (nm), one fast walking sequence (fq) and one slow walking sequence (fs). The accuracy obtained by the three classifiers has been shown in the figure.

Discussion
Our model is able to identify the person despite the speed the person is walking. This feature is useful particularly in surveillance fields when security measures have to be taken to stop crimes like robberies so that while fleeing the person is identified. Highest accuracy of 99% has been achieved by SVM in this regard (Table 4 and Fig. 10).

3.3 *Cloth Invariance*

Experimental Condition
CASIA-B data set was used for this instance, and the training set consisted of five normal sequences (nm), one sequence with bag (bg) and one sequence with cloth (cl). Validation was with one normal sequence (nm), one sequence with bag (bg) and one sequence with cloth (cl). The accuracy obtained by the three classifiers has been shown in the figure. View angle was kept constant here that is each sequence, both in the training and validation set, was of the same view angle.

Table 4 Metrics obtained showing speed invariance

Model	Accuracy	Precision	Recall	F1-Score	Support
SVM	1.00	1.00	0.99	0.99	459
ANN	1.00	0.99	1.00	0.99	459
XGBoost	0.97	0.96	0.97	0.96	459

Fig. 8 Different clothing conditions of subjects with or without bag

Discussion
Our model is able to identify the person in spite of what the person is wearing (Figure 8). He or she can have a bag or a coat but can still be recognized when the camera is set a certain view angle. Above 90% accuracy has been achieved with respect to every view angle with SVM performing better in almost all the cases (Table 8 and Fig. 11).

3.4 *Pose Invariance*

Experimental Condition
CASIA-B data set was used for this instance, and the two different experiments to test crossview and multiview invariance were carried out. For crossview, the training features were obtained at a particular angle and validated for another angle as shown in table. For multiview, the training features were obtained at each of the view angles together. Validation was then done for each angle separately. Five normal sequences (nm) were used in training sample, and one normal sequence was used in the test sample (Table 5).

Discussion
Our model is able to identify the person with regard to different views the camera may have been placed at (Fig. 9). Highest accuracy for multiview was obtained at 0° and 180° (Table 6 and Fig. 12) while in crossview the model performed the best when trained at 72° and validated at 90° (Table 7 and Fig. 13).

Table 5 Metrics obtained showing cloth invariance

Model	View angle	0	18	36	54	72	90	108	126	144	162	180
SVM	Accuracy	0.95	0.94	0.95	0.96	0.94	0.94	0.91	0.92	0.94	0.93	0.96
	Precision	0.96	0.96	0.97	0.97	0.96	0.96	0.93	0.94	0.95	0.95	0.97
	Recall	0.96	0.96	0.96	0.96	0.94	0.94	0.91	0.92	0.94	0.93	0.96
	F1	0.95	0.94	0.96	0.96	0.94	0.94	0.91	0.92	0.94	0.93	0.96
	Support	357	360	360	357	360	360	360	358	359	356	360
ANN	Accuracy	0.94	0.90	0.92	0.93	0.90	0.91	0.90	0.94	0.93	0.92	0.91
	Precision	0.95	0.92	0.94	0.94	0.92	0.93	0.92	0.96	0.95	0.94	0.93
	Recall	0.95	0.92	0.93	0.93	0.94	0.95	0.91	0.94	0.94	0.93	0.93
	F1	0.95	0.92	0.93	0.93	0.92	0.93	0.91	0.94	0.94	0.93	0.93
	Support	357	360	360	357	360	360	360	358	359	356	360
XGBoost	Accuracy	0.83	0.85	0.93	0.93	0.84	0.90	0.85	0.88	0.89	0.90	0.89
	Precision	0.84	0.86	0.92	0.94	0.85	0.92	0.85	0.89	0.90	0.91	0.90
	Recall	0.83	0.87	0.91	0.93	0.86	0.93	0.86	0.90	0.91	0.90	0.89
	F1	0.83	0.86	0.91	0.93	0.85	0.92	0.85	0.89	0.90	0.90	0.90
	Support	357	360	360	357	360	360	360	358	359	356	360

Fig. 9 Poses captured of subjects from different angles

4 Comparison with Existing Approaches

A true comparison for each of the criteria with existing approaches is difficult since almost all approaches have different experimental conditions and have worked on a subset of the criteria which has been presented here. It has been ensured that the data sets used in the comparison are same with CASIA-B used in cloth invariant and view invariant and CASIA-C used in speed invariant detection.The accuracy values showed in Table 8 represent the average accuracy values carried over the different experiment conditions so that uniformity is maintained. We observe that out method almost outperforms each of them except for view invariance where deep learning-based methods [25–30] have the advantage but this approach loses in terms of longer training time and more data required for such instances.

Table 6 Metrics obtained showing multiview invariance

Model	View angle	0	18	36	54	72	90	108	126	144	162	180
SVM	Accuracy	1.00	0.99	0.77	0.67	0.65	0.70	0.75	0.74	0.70	0.99	1.00
	Precision	0.96	0.96	0.77	0.67	0.66	0.69	0.76	0.75	0.71	0.98	0.99
	Recall	0.96	0.96	0.76	0.68	0.67	0.70	0.77	0.76	0.72	0.99	0.98
	F1	0.96	0.96	0.77	0.67	0.66	0.69	0.75	0.75	0.70	0.98	0.98
	Support	117	120	120	120	120	120	120	118	120	118	120
ANN	Accuracy	0.90	0.96	0.79	0.68	0.70	0.76	0.80	0.66	0.68	0.96	0.99
	Precision	0.89	0.95	0.78	0.69	0.71	0.75	0.81	0.67	0.69	0.95	0.98
	Recall	0.88	0.94	0.79	0.68	0.72	0.76	0.82	0.68	0.70	0.96	0.99
	F1	0.98	0.94	0.78	0.68	0.71	0.75	0.81	0.68	0.70	0.95	0.98
	Support	117	120	120	120	120	120	120	118	120	118	120
XGBoost	Accuracy	0.88	0.85	0.70	0.75	0.76	0.70	0.75	0.76	0.74	0.70	0.67
	Precision	0.85	0.86	0.71	0.73	0.72	0.72	0.77	0.78	0.76	0.71	0.68
	Recall	0.83	0.87	0.71	0.73	0.74	0.73	0.76	0.80	0.77	0.72	0.69
	F1	0.83	0.86	0.71	0.73	0.72	0.72	0.75	0.79	0.76	0.71	0.68
	Support	117	120	120	120	120	120	120	118	120	118	120

Table 7 Metrics obtained showing crossview invariance

Model	Training	Validation	Accuracy	Precision	Recall	F1 Score	Support
SVM	90	108	0.67	0.68	0.66	0.67	120
	72	90	0.77	0.78	0.77	0.76	119
	162	180	0.67	0.69	0.68	0.68	120
ANN	90	108	0.70	0.69	0.71	0.70	120
	72	90	0.76	0.76	0.76	0.76	120
	162	180	0.66	0.66	0.64	0.65	119
XGBoost	90	108	0.66	0.67	0.68	0.67	120
	72	90	0.70	0.72	0.73	0.72	120
	162	180	0.70	0.71	0.72	0.72	119

Table 8 Comparison with state-of-the-art models

	Methods	Accuracy
Speed	WBP [21]	91.66
	FDI +2DLDA [22]	89.0
	Proposed method	99
Cloth	GEI+part-based [23]	85.2
	Golden ratio segmentation [24]	93.14
	Proposed method	96
View	Stacked autoencoder [25] (Multiview)	97.58
	Proposed method	79.00
	Stacked autoencoder [25] (Crossview)	63.90
	Proposed method	67.00

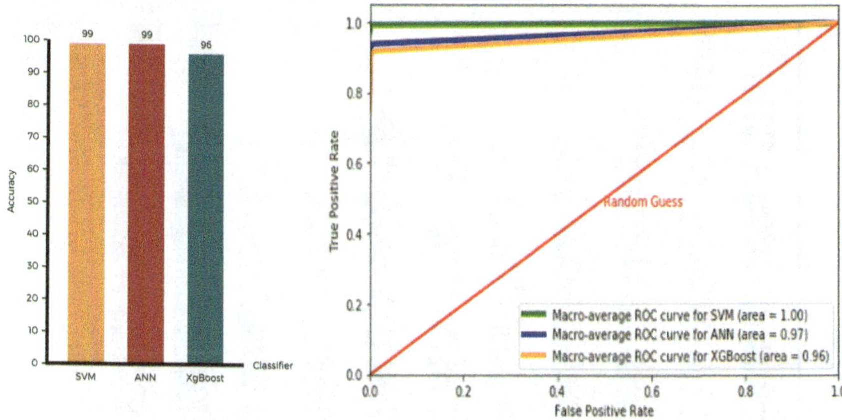

Fig. 10 Comparison of accuracy with AUC curve when subject moves with different speeds

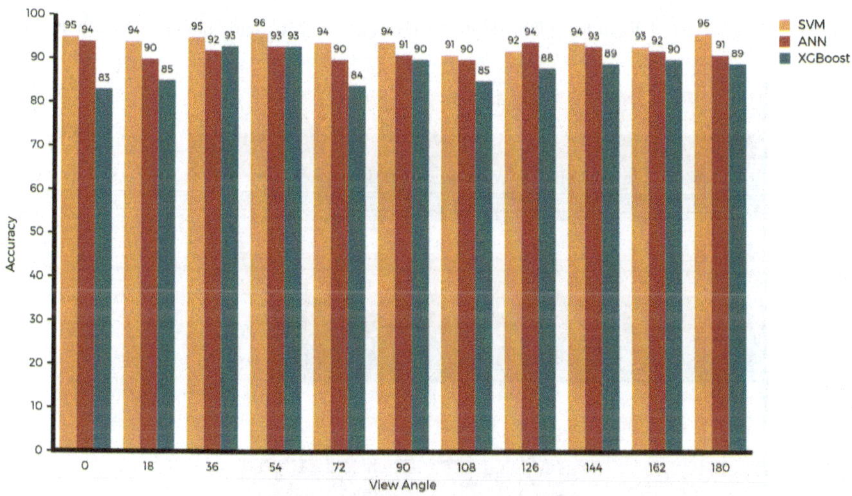

Fig. 11 Comparison of accuracy when view angle is kept same and clothing conditions changed

5 Conclusion and Future Scope

In this paper, high accuracy is achieved with normal gait and cloth invariant gait. The accuracy associated with crossview invariant gait needs to be improved. The gait recognition-based person identification becomes challenging in real time due to cloth, different view angle and posture of human. The challenge involves in extracting the correct gait features of different person which is invariant for all conditions. We have adopted technique which is invariant gait extraction. Most of the models used to consider gait from sagittal plane not from other transverse and coronal plane moment. Our algorithm is robust and works for all the view and is mapped onto single view image.

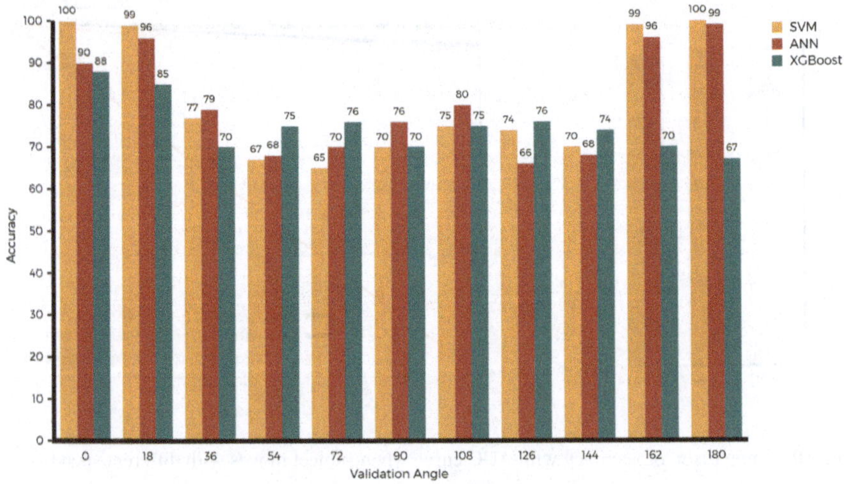

Fig. 12 Comparison of accuracy when subject is trained at 0–180° and validated at each angle

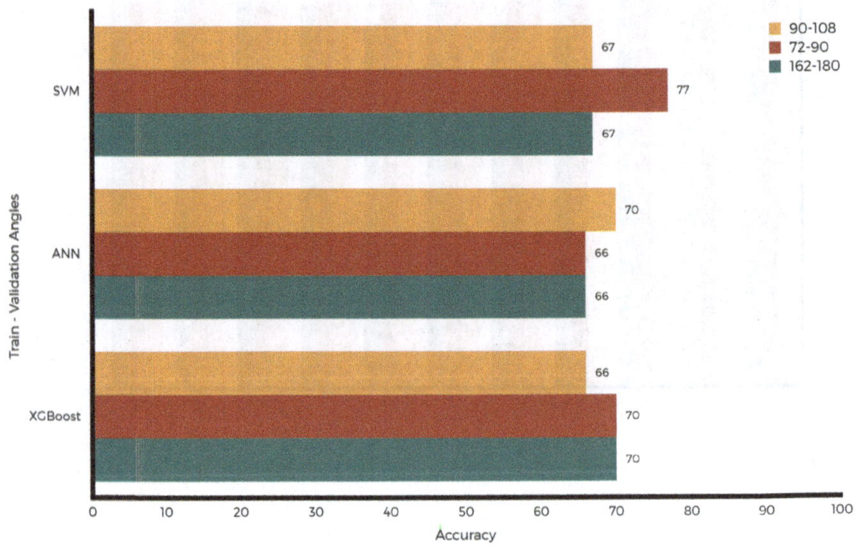

Fig. 13 Comparison of accuracy when subject is trained at a specific angle and validated at a different angle

6 Declarations

6.1 Funding

This work is partially funded by 1. SERB, DST, Government of India, for project under the schema of Early Career Award (ECR) with file no : ECR/2018/000203 dated on 04/06/2019. to Dr. Vijay Bhaskar Semwal and 2. Uttarakhand Technical University, Dehradun, for providing CRS scheme project to Mr. Vishwanath Bijalwan under TEQIP-III.

Acknowledgements The author(s) would like to thank all the participants who have allowed us to capture the data using wearable device. Special thanks to human motion capturing and analysis unit of MANIT Bhopal and Motion Capturing Sensor Laboratory of Institute of Technology Gopeshwar, India for providing opportunity to collect data and providing the basic computing facility. The data set is also available publicly for research purpose. One can download form here:Http://. We also would like to express our thanks to SERB , DST, Government of India, for funding project under schema of Early Career Award (ECR), DST No: ECR/2018/000203 dated on 04/06/2019.

Conflict of interest The author(s) declare that they have no conflict of interest.

References

1. Semwal, V. B., Raj, M., & Nandi, G. C. (2015). Biometric gait identification based on a multilayer perceptron. *Robotics and Autonomous Systems, 65*, 65–75.
2. Connor, P., & Ross, A. (2018). Biometric recognition by gait: A survey of modalities and features. *Computer Vision and Image Understanding, 167*, 1–27.
3. Yu, S., Tan, T., Huang, K., Jia, K., & Wu, X. (2009). A study on gait-based gender classification. *IEEE Transactions on Image Processing, 18*(8), 1905–1910.
4. Semwal, V. B., & Nandi, G. C. (2015). Toward developing a computational model for bipedal push recovery-A brief. *Sensors Journal, IEEE, 15*(4), 2021–2022.
5. Semwal, V. B., & Nandi, G. C. (2016). Generation of joint trajectories using hybrid automate-based model: A rocking block-based approach. *IEEE Sensors Journal, 16*(14), 5805–5816.
6. Semwal, V. B., Kumar, C., Mishra, P. K., & Nandi, G. C. (2018). Design of vector field for different subphases of gait and regeneration of gait pattern. *IEEE Transactions on Automation Science and Engineering 15*(1), 104–110 (2018)
7. Semwal, V. B., Bhushan, A., & Nandi, G. C. (2013) Study of humanoid Push recovery based on experiments. In *2013 International Conference on Control, Automation, Robotics and Embedded Systems (CARE)*. IEEE (2013)
8. Nandi, G. C., Semwal, V. B., Raj, M., & Jindal, A. (2016) Modeling bipedal locomotion trajectories using hybrid automata. In *Region 10 Conference (TENCON), 2016 IEEE*. IEEE, 2016.
9. Semwal, V. B., Singha, J., Sharma, P. K., Chauhan, A., & Behera, B. (2017). An optimized feature selection technique based on incremental feature analysis for bio-metric gait data classification. *Multimedia tools and applications, 76*(22), 24457–24475.
10. Li, W., Jay Kuo, C.-C., & Peng, J. (2018). Gait recognition via GEI subspace projections and collaborative representation classification. *Neurocomputing, 275*, 1932–1945.

11. Huan, T. T., Van Kien, Cao, Anh, H. P. H., & Nam, N. T. (2018). Adaptive gait generation for humanoid robot using evolutionary neural model optimized with modified differential evolution technique. *Neurocomputing*, *320*, 112–120.
12. Mahfouf, Z., Merouani, H. F. Bouchrika, I., & Harrati, N. (2018). Investigating the use of motion-based features from optical flow for gait recognition. *Neurocomputing, 283*.
13. Zheng, S., Huang, K., & Tan, T. (2011). Evaluation framework on translation-invariant representation for cumulative foot pressure image.
14. Semwal, V. B. (2017). Data driven computational model for bipedal walking and push recovery. arXiv preprint arXiv:1710.06548 (2017).
15. Semwal, V. B., Mondal, K., & Nandi, G. C. (2017). Robust and accurate feature selection for humanoid push recovery and classification: deep learning approach. *Neural Computing and Applications*, *28*(3), 565–574.
16. Semwal, V. B., Katiyar, S. A., Chakraborty, P., & Nandi, G. C. (2013). Biped model based on human Gait pattern parameters for sagittal plane movement. In *International Conference on Control* (p. 2013). IEEE: Automation, Robotics and Embedded Systems (CARE).
17. Semwal, V. B., Katiyar, S. A., Chakraborty, R., & Nandi, G. C. (2015). Biologically-inspired push recovery capable bipedal locomotion modeling through hybrid automata. *Robotics and Autonomous Systems*, *70*, 181–190.
18. Wang, L., Tan, T., Ning, H., & Weiming, H. (2004). Fusion of static and dynamic body biometrics for gait recognition. *IEEE Transactions on Circuits and Systems for Video*, *14*(12), 149–158.
19. Zheng, S., Zhang, J., Huang, K., He, R., & Tan, T. (2011). Robust View Transformation Model for Gait Recognition.
20. Mahfouf, Z., Merouani, H. F., Bouchrika, I., & Harrati, N. (2018). Investigating the use of motion-based features from optical flow for gait recognition. *Neurocomputing*, *283*(2018),
21. Kusakunniran, W., Wu, Q., Li, H., & Zhang, J. (2009). Automatic gait recognition using weighted binary pattern on video. In *2009 Sixth IEEE International Conference on Advanced Video and Signal Based Surveillance*, pp. 49–54.
22. Chen, C., Liang, J., Zhao, H., Hu, H., & Tian, J. (2009). Frame difference energy image for gait recognition with incomplete silhouettes. *Pattern Recognition Letters*, *30*(11), 977–984.
23. Li, N., Xu, Y., & Yang, X. (2010) Part-based human gait identification under clothing and carrying condition variations. In: *Proceeding of the Ninth International Conference on Machine Learning and Cybernetics*, pp. 268–273
24. Liang, Y., Li, C. T., Guan, Y., & Hu, Y. (2016). Gait recognition based on the golden ratio. *EURASIP Journal on Image and Video Processing*, *2016*(1), 22.
25. Yu, S., Chen, H., Wang, Q., Shen, L., & Huang, Y. Invariant feature extraction for gait recognition using only one uniform model. *Neurocomputing*.
26. Semwal, V. B., Gaud, N., & Nandi, G. C. (2019). Human gait state prediction using cellular automata and classification using ELM. Machine intelligence and signal analysis. *Springer, Singapore*, *2019*, 135–145.
27. Patil, P., Kumar, K. S., Gaud, N., & Semwal, V. B. (2019). Clinical human gait classification: extreme learning machine approach. *1st International Conference on Advances in Science* (p. 2019). IEEE: Engineering and Robotics Technology (ICASERT).
28. Raj, M., Semwal, V. B., & Nandi, G. C. (2018). Bidirectional association of joint angle trajectories for humanoid locomotion: the restricted Boltzmann machine approach. *Neural Computing and Applications*, *30*(6), 1747–1755.
29. Gupta, A., & Semwal, V. B. (2020). *Multiple task human gait analysis and identification: Ensemble learning approach*. Springer Plus: Emotion and Information Processing.
30. Kumar, M., Gautam, P., & Semwal, V. B. (2020). A Review of computational model for bipedal robot walking using gait analysis. In *2020 International Conference on Emerging Smart Computing and Informatics (ESCI)*. IEEE.

Application of Machine Learning in Industry 4.0

Mahendra Kumar Gourisaria, Rakshit Agrawal, Harshvardhan GM, Manjusha Pandey, and Siddharth Swarup Rautaray

Abstract The fourth industrial revolution, Industry 4.0, is concerned with the ongoing changes and advances in the way of manufacturing and processes involved for the production of goods and their efficiency which is increasing rapidly with the advancement in the machine learning techniques and other technologies. Industry 4.0 encapsulates future industry development trends and techniques so as to achieve more advanced and intelligent manufacturing processes. Making use of smart, advanced, and automated machines, products are manufactured in more efficient, customized, and smarter way. It led a way of smart production lines which improved flexibility for manufacturers to experiment with different product designs and configurations—a strategic initiative introduced by the German government. It is a step toward transformation of industrial manufacturing through digitalization and exploitation of potentials of new approaches and technologies. With the access to usage data from the machine and product's feedback, overall efficiency of production is improved. The effect of digitization and advancement in technology strives to make life simpler.

Keywords Industry 4.0 · Machine learning · Digitalization · Advancement in technology · Big data analysis · Augmented reality (AR) · Additive manufacturing · Cloud computing · Cyber security · Internet of things (IoT)

M. K. Gourisaria (✉) · R. Agrawal · Harshvardhan GM · M. Pandey · S. S. Rautaray
School of Computer Engineering, KIIT (Deemed to be University), Bhubaneswar 751024, Odisha, India
e-mail: mkgourisaria2010@gmail.com

R. Agrawal
e-mail: rakmak3456@gmail.com

Harshvardhan GM
e-mail: harrshvardhan@gmail.com

M. Pandey
e-mail: manjushafcs@kiit.ac.in

S. S. Rautaray
e-mail: siddharthfcs@kiit.ac.in

M. Pandey and S. S. Rautaray (eds.), *Machine Learning: Theoretical Foundations and Practical Applications*, Studies in Big Data 87,
https://doi.org/10.1007/978-981-33-6518-6_4

1 Industry 4.0

Industry 4.0 is bag of nine different technical concepts which are big data analysis, augmented reality (AR), additive manufacturing, cloud computing, cyber security, Internet of things (IoT), system integration, autonomous robots, and simulation that has changed the traditional way of production drastically. Figure 1 shows the transition of industry from 1800 s to ongoing era of Industry 4.0. Industry 4.0 is making a digital transformation sector using smart manufacturing, resulting in cyber physical systems.

The machinery and other products are becoming more cloud based and automated so as to take decisions on their own and so reducing the human labor work. With the help of machine learning techniques and algorithms, machines are showing drastic and efficient changes in every department of the industry. Based upon the need and required type of product, different machine learning techniques are applied so as to maximize the efficiency, accuracy, and other factors which makes a product smart and efficient. Interconnected automation machines and cloud-based systems help in taking effective and agile manufacturing decisions which lead to smarter way of productivity. Machine learning is itself a vast topic which covers different techniques and algorithms which makes automated systems. The chapter covers different techniques of machine learning, namely regressors, classifiers, clustering, natural language process, reinforcement learning, and some techniques of deep learning namely: artificial neural networks (ANN), convolutional neural networks (CNN), autoencoders, and Boltzmann machine along with their applications. These techniques, however, fall under different sub heading of machine learning. Regressor and classifier techniques are supervised learning which require a dataset for the training process, and then, the model can be applied for decision making, whereas clustering falls under unsupervised learning which is opposite of supervised learning. Classifiers are used in different fields of industry. Clustering is broadly classified into two types further: K-means clustering and hierarchical clustering. Deep learning has many practical applications in Industry 4.0 which plays a vital role in medical industries for getting accurate diagnostic results. Similar to machine learning, deep

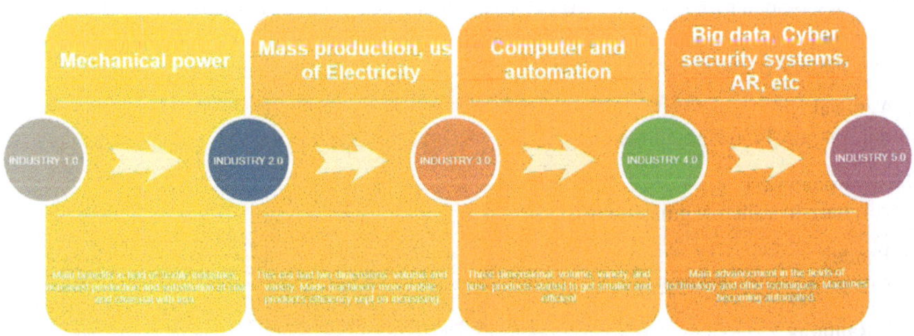

Fig. 1 Industrial revolution from 1.0– 4.0 with their main advantages

learning too has sub categories: supervised and unsupervised. Boltzmann machines and autoencoders are a part of unsupervised learning, whereas CNN and ANN fall under the category of supervised learning.

Industry 4.0 has already changed the way of production of goods and has made them efficient, customized, and smarter with the help of different machine learning techniques which leads to the beginning of the future of many developing industries that can create new business models, cloud-based systems, etc. This will create opportunities in various departments of industrial sector leading the expansion of industry.

Industry 4.0 is a subspace of fourth industrial revolution in the field of manufacturing which mainly deals with industries [1]. Industry 4.0 is the composition of cyber physical systems, Internet of things (IoT), Internet of systems, cognitive computing, cloud computing, artificial intelligence, etc. Use of smart machines, industries are manufacturing goods in a more efficient way and getting better and better with less power consumption and pollution. In this revolution, the machines are connected with one another digitally and make and share information making it more efficient and ever faster than before [2]. Advancement in the technical field has led to a drastic increase in the productivity of goods in industries since from the beginning of industrial revolution. Factories were based on the steam-powered engines in the nineteenth century, with the introduction of electricity, it led to the mass production of good in early twentieth century and after this industries became autonomous with the fourth revolution: Industry 4.0. The aim of the revolution is to bring autonomous decision-making processes in the machines and data exchange in real time view. With the help of machine learning algorithms and techniques, machines are effectively making decisions for processing. Wireless connectivity of machines and with the help of autonomous decision making ability of machines, it will provide faster response time and upstanding results.

Industry 4.0 came with many advantages for the manufacturers and advancement in the technology [3]. It increased the productivity in a smarter way with interconnected automation making agile and effective manufacturing decisions to meet the target. It led a way of smart production lines which improved flexibility for manufacturers to experiment with different product designs and configuration. With the access to usage data from the machines and product's feedbacks, product's quality can be improved. Industry 4.0 offers manufacturers to provide wide range of customized products depending upon customer's needs.

In this chapter, we discuss how Industry 4.0 came into being with all its different aspects, and with it, we describe the taxonomy of modern-day machine learning techniques that are used in Industry 4.0 with their applications. These techniques include regression, classification, clustering, natural language processing (NLP), reinforcement learning, and deep learning.

2 Evolution of Industry 1.0–4.0

The change in the methods of production of goods in industries from manual manufacturing to smart and automated manufacturing is referred as industrial evolution. In the late eighteenth and nineteenth century, goods like cloths, food, shelters, and weapons were manufactured by humans. With the introduction of Industry 1.0, there were drastic changes in the methods for the production of goods and operations with the evolution developed rapidly. Figure 2 shows a basic overview of the evolution of the Industry 1.0–4.0.

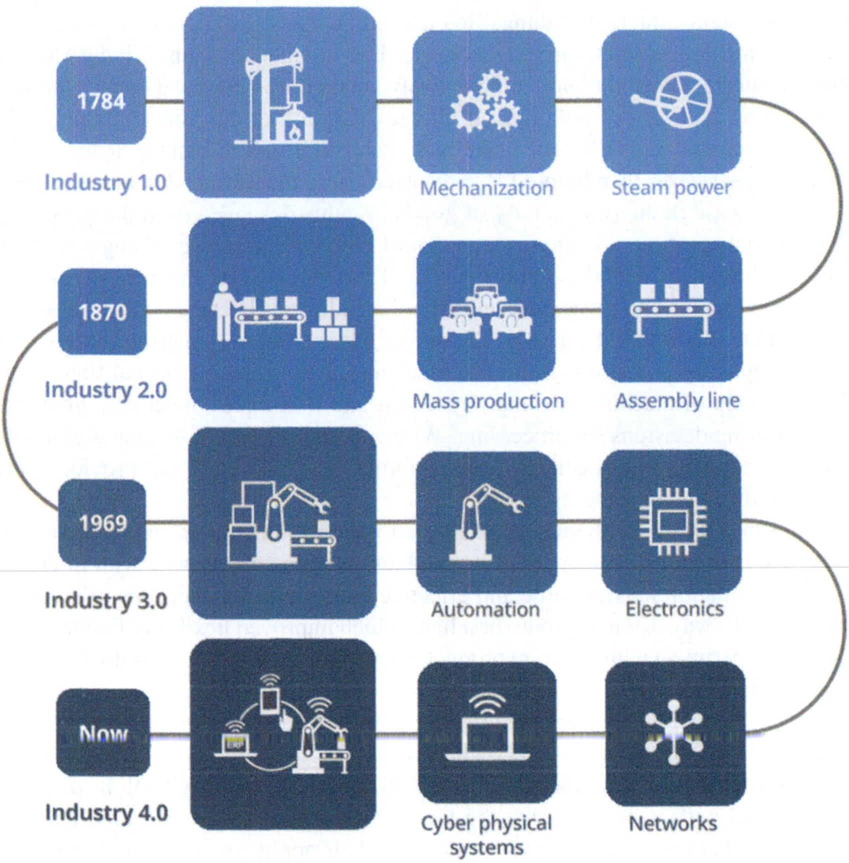

Fig. 2 Industry 1.0–4.0

2.1 *Industry 1.0*

The first industrial evolution happened to be in mid eighteenth century. This brought the activities of human from focusing on agricultural to the industrial society. In this era, demand of products was one dimensional, i.e., product volume. The evolution mainly aimed at manual labor manufacturing performed by people around with the help of animal labor and making use of more optimized form of labor performed by people by taking the help of water and steam-based power engines and other types of machine tools [4]. The main benefits of the First Industrial Revolution are provided: In textile industry, there was a drastic increase in the production and efficiency with the use of steam powered cotton spinning machines. The cost of iron production in this era decreased because of the substitution of coal as a fuel on charcoal.

2.2 *Industry 2.0*

The second era of industrial revolution began from nineteenth century and ended by early twentieth century. Industry 2.0 is built on the ideas and inventions of the first industrial revolution. With the increasing demand of goods in Industry 1.0, introduction of steel and use of electricity came into picture. This era came with a two dimension: product volume and variety. Major technological innovations included electricity, electronic and advanced mechanical devices, and cars. The use of electricity enabled manufacturers to increase their efficiency and helped to make factory machineries more mobile. The concept of mass production so as to increase the productivity came during this phase [5].

2.3 *Industry 3.0*

The third revolution in industries began in twentieth century, with the approach of digitization of the machines. This revolution had a great impact, mainly to the electronic industries. The demands of product in the third revolution increased to three dimensions: product volume, variety, and delivery time. Most of the electronic product's architecture changed from integral to modular, which also reduced the average life span of the goods. This era came with the advancement in mechanical calculators, computers, and development of communication techniques. The products were getting smaller and advanced which made them a part of our daily lives with the use of computer and communication technologies in the process of production.

2.4 *Industry 4.0*

Industry 4.0 is the fourth and ongoing revolution that will make use of smart machines and the 9 pillars as shown in Fig. 1. Industry 4.0 is an initiative with innovation and advancement in the field of technologies. Researchers, industries, and even government are attracted toward this. The fourth generation of industrial revolution tends to make digital transformation unite with smart manufacturing, resulting in cyber physical systems.

3 Nine Pillars of Industry 4.0

Industry 4.0 is making use of advanced digital technologies for manufacturing. This is leading to greater efficiencies and has changed the traditional way of production and the relationship among the suppliers, producers, and customers. The revolution has also changed the relation between humans and machines. Figure 3 shows nine technologies which are the building blocks for the fourth revolution of this era.

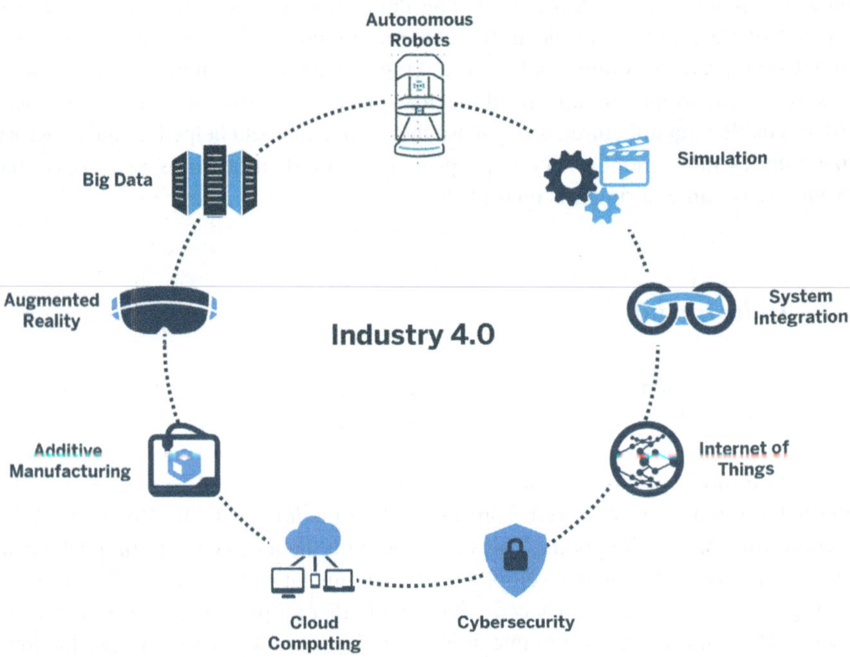

Fig. 3 9 pillars of Industry 4.0

3.1 Big Data and Analytics

In the context of Industry 4.0, improvement in the level of data and advancement in the field of technology accelerates firm's advantage in increasing productivity, innovation of products, and overall competition [6]. With the help of decision makers, big data projects led to work out on challenges at organizational level with appropriate monitoring, measuring, and managing. Big data has further subtypes which are data as a tool (traditional way of manufacturing chain continued by existing capabilities), data as an industry (new upcoming ventures and development of software for working with big data), and data as a strategy (making data resources by developing new innovation business models). Big data analytics involve some processes to analyze the large datasets provided by which some information in the terms of preferences, algorithms based on correlation, trends, and other information of the customers. This big data analytics with the help of machine learning techniques can be used in various areas for the development of the ventures, companies, product quality, etc., and those area scan be fault prediction so as to reduce error rate, the predictive algorithms that can reduce the harm before damages happen. The data can be extracted from the system itself for the immediate insights, and prediction of outcomes as well as decision making can be made simple. The capability of managing and analyzing big data gives a competitive advantage to the firms and benefits them with their operations, marketing, improvement in customer's feedback and experience, and a lot more.

A real-time instance of big data analytics can be shown in Infineon technologies: A semiconductor manufacturer has reduced product failure with help of machine learning algorithms and correlating the data captured of the single chip in the testing phase. In this way, the company tends to differentiate between different patterns that help in removing the faulty manufactured chips early in the production phase, hence improving the production quality.

3.2 Autonomous Robots

Many industries manufacture robots which are trained (using machine learning) to handle complex problems and proceed accordingly. With the advancement in the field of technology, manufacturers are trying to make more advance robots with advance abilities to perform task and handle more complex situations and give accurate and efficient results. In industries, these advance robots can be used for various purposes and in different areas like production, logical problem solving, and distribution activities. Robots are designed and programmed with the help of machine learning algorithms and techniques in such a way that they can learn from humans and perform the same task more efficiently and even in lesser time. These new robots will be less costly and will have a wide range of capabilities in comparison to the old manufactured robots. Bahrin et al. [7] discussed several technologies introduced by different companies in the field of robotics technology such as Kuka LBR IIWA. These robots having these technology programmed in them can now have the ability to learn from human and do the work in an optimized way.

3.3 Simulation

Industry 4.0 provides an advanced simulation tool which plays an important and supportive role in the activities related to production and promotes sustainable manufacturing environment. These simulation tools with the help of real-time data to form an virtual model out of it which will include machines, products, etc. By this, manufacturers and operators will be allowed to test and optimize the machine for the production of goods in the virtual model before applying them in real time, reducing the optimizing time and increasing the product's quality. The virtual model can be used for dynamic research for production system with the help of real time datasets, hence optimizing the overall operation.

3.4 Internet of Things

The machines which are manufactured and used in the industries after the fourth industrial revolution make use of embedded computing. Machine's sensors and other machines are connected and share information as per the requirements so as to boost the efficiency and reduce the setup time. IoT refers to the technological revolution by providing solution for advance computations, analytics, and many more by providing a cloud-based system. With the interlinking of machines and other devices with one another, allows them to interact and with more centralized controllers. IoT connects to Internet (cloud-based system) with the data obtained from the machines, different departments, customers (feedbacks), etc. The system makes the proceeding according to the data collected and provides the results. By IoT and cloud-based system, not only business operations, but also other companies' operations in different fields become more agile and integrated. IoT helps in efficient decision making and it also enables real-time response (necessary for efficient work).

3.5 Cloud

Cloud computing have a variety of advantages in the field of industries and other various firms like helping SCs to automize, integrate, and also facilitate departments and administration. Companies use cloud-based software for different enterprise and analytical applications which helps in sharing big data in the company boundaries. In cloud computing, there are namely three models: Software as a service (SaaS), platform as a service (PaaS), and infrastructure as a service (IaaS). In SaaS, access depends on the customer purchase. PaaS is one in which customers allowed to use the applications provided on the cloud. Iaas offers basic activities on the cloud. The most popular example for the model can be Google Drive by Google.

3.6 System Integration

In system integration, there are further two types: horizontal integration and vertical integration. In the era of smart factories and use of cloud-based systems, horizontal integration helps different machineries, IoT devices, work of different departments, etc., to process together to provide a better output. On the other hand, vertical integration makes sure that data from different departments, machines, etc., is used at higher organizational level for making marketing, staffing, and other necessary decisions. Data flows freely and transparently up and down these layers so that both strategic and tactical decisions can be according to the data. This data can be present on the cloud-based systems or the big data collected from the machines and other IoT devices or both. The coordination of horizontal and vertical integration forms the framework of the smart factory.

3.7 Additive Manufacturing

The fourth industrial revolution focuses on integration of intelligent and efficient production systems and advanced informational technologies. Additive manufacturing is an important pillar of Industry 4.0 and the most widespread technologies for the production of small batches of customized lightweight design with ease of construction. Nowadays, it is possible to create flexible, automated products, implementing multiplex technologies, interlinking physical components, and information systems. With additive manufacturing, 3D printing and rapid prototyping have revolutionized the fields of production. Advanced parts are manufactured using computer aided design software.

3.8 Augmented Reality

Augmented reality is an essential technology in Industry 4.0, provides a replica of the physical world in terms of virtual environment [8]. Currently, annual growth in AR industry is expected to be 74% between 2018 and 2025 [9]. In industry, AR can be used for customized designs and manufacturing applications, different assembly operations, and many more. AR proves to solve tasks in efficient, less time consuming, and negligible error rate, however, it depends upon the complexity of the task as well the nature of it, and so, selection of task for AR is a important step before implementation. AR systems have some basic components like visualization technology, camera, tracking devices, and user interface. This technology increases human machine interaction with more ease, tasks controlled by remotes. For controlling the required tasks, AR provides motion controls using tracking sensors for its users. World's first augmented reality glasses were launched by Google and named as Google glass which are adjusted according to the human eyes [10].

3.9 Cyber Security

Cyber security is a major issue for all the companies which can have a worse impact on business or may affect the efficiency of the company because of the harmful cyber-attacks. However, there are solutions for these attacks which are implemented by many companies so as to protect their cloud-based systems, data, and other crucial information.

4 Machine Learning

Around 1960s, the term machine learning was first introduced by Arthur Samuel. Machine learning mainly deals with pattern classification. Machine learning is based on the theory that computers can be programmed in such a manner that they can learn different tasks without being programmed to do the specific task. Advancement in the field of machine learning kept going on, advance pattern recognition techniques, strategies, algorithms, etc. Machine learning is a computer program from which machine learns the pattern, and relationship between the features present in the dataset is provided. Machine learning is a subfield of artificial intelligence, in which systems are provided with different datasets and various algorithm techniques that helps in creating the relationship in between the data, analyze, and interpret the type of data on their own. In machine learning iterative aspect is crucial because when the model (programmed system) is exposed to a set of new datasets, they should be capable of adapting and forming relations for an accurate result. Different models have different way of making relations and algorithms for the same datasets.

Over the years, development in the field of machine learning has drastically increased. Now, machine learning has ability to automatically implement complex mathematical calculations to big data in lesser time with less error rate and more efficientcy. Machine learning is a field that is been used by many industries as well as companies for better results and efficiency, used in financial services, government, medical field, etc. Machine learning is a wide field having various techniques and algorithms that can be implemented on the datasets and obtain best results out of it. Figure 4 shows different algorithms present in machine learning.

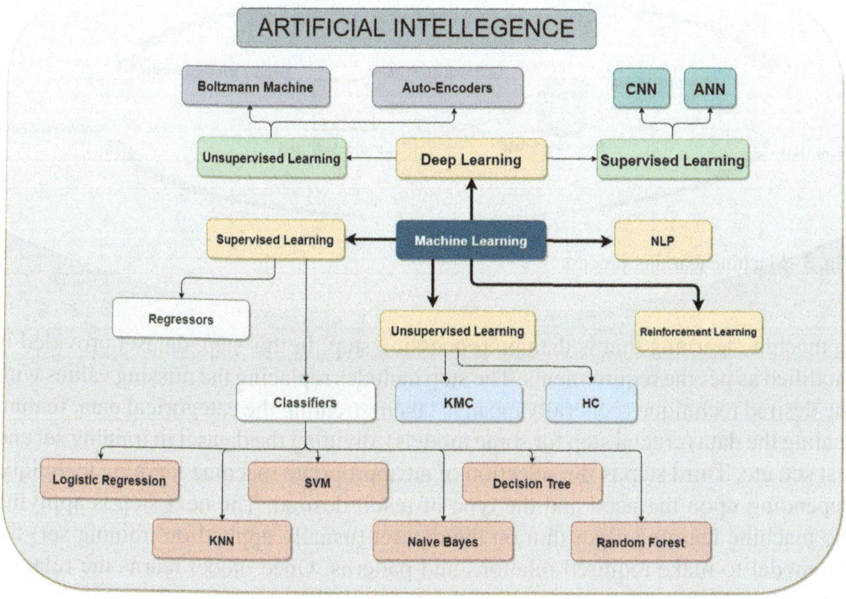

Fig. 4 Flowchart for machine learning

Machine learning being a vast field, mainly divided into five types: Deep learning, supervised learning, unsupervised learning, reinforcement learning, and natural processing language (NLP). Apart from these four subtypes of machine learning, there are many more which can be implemented on specific datasets.

In machine learning, supervised learning is a learning in which a set of labeled data (known instances) is provided for the model to make acceptable relations between the features. From these relations, a model is able to predict the results for the new datasets on the basis of the relations developed for the existing datasets. Supervised learning is a powerful tool for classification and data processing. On the other hand, unsupervised learning is type of machine learning which is used to make relations from the datasets consisting instances without labeled responses. It can be termed as the training of machine using the information that is neither classified nor labeled and allowing the desired algorithm to make relations without any prior knowledge. The main task of unsupervised learning is to organize the information in groups according to the similarities, patterns, and relation between the features without any guidance (program).

Machine learning mainly involves five processes for getting the results. First step is the collection of the data. Providing authentic data is prerequisite condition for applying machine learning model. Depending upon the machine learning algorithm desired, required types of datasets are required. In supervised learning, labeled datasets are required, and in case of unsupervised learning, there is no need for providing datasets having labeled responses. The second step is an important step

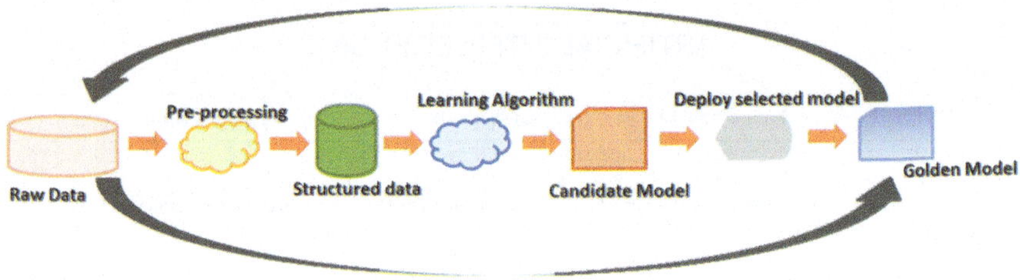

Fig. 5 Machine learning process

in machine learning that is data preprocessing step. In this step, dataset provided is modified as per the requirements. The step includes replacing the missing values with the desired technique (generally mean is taken), treating the categorical data, feature scaling the data (crucial step for some models), dividing the dataset in training set and test set, etc. Third step is the selection of an appropriate machine learning technique depending upon the need and the type of result desired. The next step is applying the machine learning algorithm on the dataset (usually applied on training set) for the model to make required relations and patterns. Once model learns the relation for the provided datasets, new data can be provided to the model for the prediction of result. Depending upon the data preprocessing and selection of machine learning algorithms, the accuracy, precision and other metrics are achieved. Figure 5 shows the order of steps followed in machine learning model.

Supervised learning is further branched in two types: Regressors and classifiers are the most common and efficient way for predicting the outcomes based on the dataset provided (labeled data) in machine learning. Regressors and classifiers have further subtypes which have different algorithms and techniques to make relations in the datasets.

4.1 Regressors

Regressors are techniques which fall under the supervised learning category of machine learning. Regression models are generally used to predict continuous values like prediction of house's cost, salary of an employee, etc. Regressors are further classified into five types: Logistic regression, polynomial regression, support vector, decision tree, and random forest regression.

A. *Logistic Regression*

It is common technique used for datasets having linear relationship in them. The prediction is based on the independent variables (input variables). The aim of the model is to find best fit line such that the results are more accurate of the training set. Depending upon the relation formed, new predictions are made on new data points.

$$Y = m + bX \tag{1}$$

In Eq. 1, Y is the dependent variable which predicts the value depending upon the independent variable X, and m and b are model coefficients.

B. *Polynomial Regression*

In this type of regression, a curved line is formed for the best fit. The curve depends upon the degree (selected as per the need) given in the function. Polynomial regression techniques are prone to be over fitted over data if the selected degree has a high value as it will start making relation including the unwanted or less necessary data. Equation 2 shows the relation between the dependent and independent variable.

$$Y = b_0 + b_1X_1 + b_2X_2^2 + \cdots + b_nX_n^n \tag{2}$$

where b_0 is the model's constant and $b_1, b_2, \ldots, b_n$ are coefficients of the independent variables $X_1, X_2, \ldots, X_3$.

C. *Support Vector Regression*

SVR is a technique in which data points are separated by a hyper plane. The model's aim is to maximize the margin in such a way that it separated the two or more groups as widest as possible. SVR is similar to the SVM technique which is used in classifiers.

D. *Decision Tree Regression*

Decision tree is a popular technique which can be used both as regressor as well as a classifier. A decision tree is made by making partitions in the datasets having similar or homogenous values. In regression, usually, mean is taken of each branch so as to avoid over-fitting.

E. *Random Forest Regression*

Random forest is a technique in which many decision trees are taken together. Because of many decision trees taking part in random forest regression, overall accuracy is increased and error rate becomes lesser and lesser. This type of regression also avoids over-fitting by making random subsets and makes small trees of these subsets.

Table 1 shows an overview of all the regressors with their pros and cons.

4.2 *Classifiers*

Machine learning classification is the process of categorizing the groups of data present in the datasets. There are six classifier algorithms which are used frequently depending upon the problems are the type of result required. Classifiers have wide

Table 1 Overview of all the regressors

Regressor	Pros	Cons
Logistic regression	Works on any size of dataset. Give information of relevance features	The assumptions linear regressor take for predicting the results
Polynomial regression	Works on both linear and nonlinear datasets	Need to choose a right polynomial degree. Over-fitting might occur
Support vector regressor	Easily adaptable. Not biased by outliers. Suitable choice for both linear and nonlinear datasets	Difficult to interpret. Feature scaling is a mandatory step
Decision tree regressor	Easy to interpret. Feature scaling is not mandatory. Can work on both linear and nonlinear problems	Poor results on too small datasets. Prone to over-fitting
Random forest regressor	It is powerful and gives accurate results	Need to choose the number of trees. Has no interpretability

variety of uses depending upon the need. Classifiers are used in different fields of industry, even used in medical industry [11–13]. These classifiers are logistic regression, K-nearest neighbor, support vector machine, Naive bayes, decision tree, and random forest.

A. *Logistic Regression Classifier*

Logistic regression recognizes the relationship and patterns between binary outcome (dependent) variable and predictor (independent) variables. If $X_1, X_2 \ldots X_n$ defines n number of independent features, and Y shows the predicted result, the equation below shows the relation between the predictor variables and p:

$$\log\left(\frac{p}{1-p}\right) = b_0 + b_1 X_1 + \cdots + b_n X_n \tag{3}$$

In the equation above b_0 is a constant and $b_1, b_2, \ldots, b_n$ are the regression coefficients of the predictor variables $X_1, X_2, \ldots, X_n$. Since the power of these regression coefficients is unity, logistic regression is termed as a linear regression.

B. *K-Nearest Neighbors Classifier*

In machine learning, one of the most used algorithms is K-nearest neighbor algorithm. It is based on instances that do not require learning phase. The element is compared to other elements using similar measures before classifying it a new element. Then its k-nearest neighbors are considered, and the class that appears more frequently among the neighbors is assigned to the element to be classified. The criterion for the classification of neighbors is the distance that separates the new element. Since KNN classifier relates data points on the bases of distance (shown in Eq. 4), it becomes a necessary task to scale the features before setting up the model.

$$d(x_i, x_j) = \sqrt{\sum_{r=1}^{n} (x_{ir} - x_{jr})^2} \tag{4}$$

C. *Support Vector Machine Classifier (SVM)*

SVM is a popular and efficient classification technique of machine learning. It works by selecting critical data points (support vectors) from all categories and separates by generating a linear function that split them as broadly as possible using support vectors with a hyper plane. The linear function aims at maximizing distance between the hyper plane and nearest data points by finding best suited hyper plane. SVM is widely used in different fields of industries [14–18].

D. *Naive Bayes Classifier*

It is a classifier which is based on probabilistic approach. It calculates a set of probability by computing frequency and combination of values in the dataset. The Bayes' theorem used in the classifier is a mathematical theorem which is used to determine conditional probability as shown in Eq. 5.

$$P(A|B) = \frac{P(A) \times P(B|A)}{P(B)} \tag{5}$$

In the above equation, $P(A|B)$ is the probability of the occurrence of event A when event B occurs, $P(A)$ is the probability of occurrence of event A, $P(B|A)$ is the probability of occurrence of event B when event A occurs, and $P(B)$ is the probability of the occurrence of B.

E. *Decision Tree Classifier*

Decision tree classifier is a reliable classification technique in machine learning. This algorithm has been very favorable with the advancement of machine learning in the field of information systems. This technique forms a tree like structure by separating the observations in branches, hence increasing the performance of the classifier and so as accuracy. A decision tree begins from root, and the node at which the tree ends is termed as leaf node. Non-leaf node shows a test or condition on a specific attribute and every branch represents the outcome of the test and leaf node depicts a decision or classification. There can be misclassification error in the classifier due to over-fitting. Since this classifier forms a tree structure, it makes easy for a better understanding and interpretations.

F. *Random Forest Classifier*

Random forest classifier is a machine learning technique that is a group of decisions tress which is constructed from random selection of records from the training set. Since random forest uses group of decision trees for the prediction, the overall performance of this classifier is found to be better than a single decision tree classifier. However, random forest classifier is more vigorous to noise than the decision

Table 2 Overview for all the classifiers

Classifier	Advantages	Disadvantages
Logistic regression	It has probabilistic approach. Tells about statistical significance of features	The assumptions that comes while making the model
K-nearest neighbors	Better understanding. It is fast and efficient	Need to pick a right value for neighbors '*k*'
Support vector machine	It has high performance. Is not sensitive to over-fitting	Not efficient over nonlinear dataset problems
Naive bayes	Efficient work on nonlinear problems. Comes with a probabilistic approach	Assumes that features are of same statistical relevance
Decision tress	It is easy to interpret. Works on linear and nonlinear problems	Under par results on too small datasets. Over-fitting is easy to occur
Random forest	It is powerful and accurate. It can be applied on linear as well as on nonlinear problems	No interpretability. More likely to have over-fitting. Need to choose the number of trees

tree classifier, present in the data provided. It has a wide variety of applications in many fields. Random forest is used in gaming industry where it recognizes human behavior and proceeds further. Microsoft Kinect is using random forest technique for human posture recognition. There are many more companies in the industry which are now using these algorithms for various functions making the games more exciting and efficient with the advancement in the field of technology and machine learning algorithms. This techniques is also used in various other fields of industry [19–22].

Table 2 shows an overview of all the classifiers discussed above.

4.3 Clustering

Clustering is a type of unsupervised learning in machine learning. Clustering divides the data points into groups on the basis of their similarities. Clustering determines the intrinsic grouping among the non-labeled datasets. Clustering techniques can be used in practical applications as well [23–28]. Depending upon the need and type of grouping required, clustering has two techniques: K-means clustering and hierarchical clustering.

A. *K-Means Clustering (*KMC*)*

K-means clustering is a common technique of machine learning. The objective of this algorithm is to group data points on the basis of their similarities together and find the unknown patterns. To achieve the objective search for a fixed number of K points (no of centroids), this is specified in the program depending upon the need. In K-means, means here refer to averaging the distance in finding the centroids.

B. *Hierarchical Clustering (HC)*

Hierarchical clustering is used to group objects in clusters based on their similarities. Agglomerative hierarchical clustering is a subtype of hierarchical clustering which is a popular technique for grouping the data. This technique makes each data point a new cluster and so forms N clusters (N data points). Then it considers two nearest data points and assigns them in single cluster, forming N-2 clusters. The above steps are repeated until only one cluster is left for finding the correct number of clusters dendrogram (type of tree diagram showing hierarchical relations between different sets).

4.4 Reinforcement Learning

Reinforcement learning can be referred as training of the machine learning model for making sequence of decisions. The model learns to achieve a set goal in an uncertain and a complex environment. The model employs a trial and error to come up with a solution for the specific task that has to be performed. If the task performed is successfully employed, machine is rewarded in some way, generally it is points, and if model faces an error in performing the task, negative or no point or some kind of penalty is awarded. By this algorithm, model tries to find best possible behavior or path (direction) that should be taken in that particular situation and learns how to perform a task by making a sequence of decisions so as to maximize the number of rewards. In reinforcement learning, an input data is provided from which the model will start. The model performs trials so as to maximize the reward and learns how to deal with the specific problem. However, reinforcement learning is not same as supervised learning. Table 3 shows the differences between reinforcement learning and supervised learning.

Reinforcement learning is divided into two subtypes: positive and negative reinforcement learning.

A. *Positive Reinforcement Learning*

It is a learning in which when an event, occurs due to some particular behavior, increase the strength or the number of rewards as well as the frequency of the behavior.

Table 3 Comparison between reinforcement and supervised learning

Reinforcement learning	Supervised learning
Decisions are made sequentially. Output is dependent on the input, and the next input depends on the output of the previous input	The decision is based on the initial input or the input given at the start (dataset provided)
Decisions are dependent, and some labels are given to sequences of dependent decisions	The decisions are independent of each other so labels are given to each decision
For example: chess game	For example: object recognition

This type of learning maximizes the performance of the model and can sustain changes for a longer period of time. But on the other hand, excess reinforcement can lead to overload of states which can diminish the results.

B. *Negative Reinforcement Learning*

It is defined as strengthening of a behavior since a negative or penalty condition is avoided by the model. This learning has its own advantages like it increases the number of behaviors and provides defiance to minimum standard performance. But has some disadvantages too like it only provides enough to meet the minimum behavior.

C. *Application of Reinforcement Learning*

Reinforcement learning can be seen as a prominent approach to be applied in the models to make them learn thing on their own. Reinforcement learning has various practical applications like it is majorly used in robotics field of industrial automation, widely used in machine learning techniques and data mining, used to create training systems that provides customized set of instructions and materials according to the task to be learned by the model. Reinforcement learning can be used even in large environments. Reinforcement learning is widely used in industrial field for training the machine to perform specific task and so reducing the labor work.

The key distinguishing factor of reinforcement learning is how the agent or the model is trained to make decisions for a specific set of problems. Instead of inspecting or analyzing the data provided, as done in supervised learning, reinforcement model interact with the environment, tries different ways to find a solution, and so maximize the rewards. Reinforcement learning is the most likely way to make a machine learn things on their own in many creative ways.

4.5 Natural Language Processing

Natural learning Processing (NLP) is widely defined as manipulation of natural language like speech, text with the help of software which are programmed to analyze the data provided accordingly. It is field that focus on the interaction of human language and computer systems. NLP comes under the field of artificial intelligence in which computer analyzes, understand, and derive meaning from the data containing human language in a smart, efficient, and useful way. NLP is used to analyze text, allowing machine to make patterns and relations to understand the way of human speaking. With the help of human-computer interaction, it allows real-world applications like automatic text summarization, sentiment analysis, topic extraction, named entity recognition, parts of speech tagging, relationship extraction, stemming, etc. NLP can be mainly used for text mining, machine translation, and automated question answering. However, language is one of the easiest things for the human's mind to understand and give response, the ambiguity of language and how they are linked together makes a difficult task for natural language processing models.

NLP algorithms have various practical applications in the fields of industry, companies, etc. It allows the respective software that are programmed to understand the type, way, etc., of text of human language. NLP is widely used for sentiment analysis which can be applied in the web and social media monitoring as it allows businesses to gain to broad public opinion on the organization and its services. Applying NLP algorithm to the digital platform can help the business of any industry to acquire the needs, opinions, and intent that users share on the web and social media. NLP is also used in email filters, stops spam-based email contents from entering into mailbox. This technique is widely used in voice recognition and information extraction which can be very useful in the industrial departments so as to improve their cloud-based data.

4.6 Deep Learning

Deep learning is a branch of machine learning which only recently gained popularity. More specifically, industries worldwide started applying deep learning techniques only after a decade into the twenty-first century. Unlike machine learning, deep learning algorithms are designed in such a way that humans are not required to specify what features pertain to a particular object for prediction. For example, machine learning cannot very accurately predict the correct object given its image, solely due to the sheer amount (say, thousands) of features that are to be learnt—which may also differ from object to object. To classify cats from dogs, the model has to learn facial features, but not quite so in classifying cars from motorcycles. Thus, deep learning makes use of *neural networks*, as we shall see in this section, to automate the feature detection process and let the computer decide what features are important and what are not. Deep learning has many practical application in the Industry 4.0 and plays a vital role in medical industries for getting accurate results [29]. Neural networks are modeled after the human brain to try and mimic the interconnection of various neurons which fire based on the inputs they receive. The artificial neural network (ANN), as we discuss in the next section, is considered to be the backbone of deep learning. In literature, ANNs are also referred to as multi-layered perceptron (MLP), which had been invented already by Rosenblatt F.F. back in 1958 [30].

The next subparts of this sections are organized as follows. In Section A, we introduce and describe the ANN. Secondly, the convolutional neural network (CNN) is described in Section B which concludes our report on the supervised class of deep learning. In Section C, we move to the unsupervised class of deep learning algorithms and introduce restricted Boltzmann machines (RBMs). Finally, in Section D, we describe autoencoders which are unsupervised as well but also referred to be *self-supervised* deep learning algorithms.

A. *The Artificial Neural Network (ANN)*

ANNs are one of the most powerful tools of machine learning in terms of regressive and classifying abilities. An ANN consists of a series interconnected nodes which are organized layer-wise, as shown in Fig. 6. Connections between nodes are defined as weights which are adjusted every time the input data is passed to and fro through the network. The input is fed to the network at the input layer (left-most layer) in scaled form which passes through the nodes and multiplied by weights between the connections of two nodes in consecutive layers. In mathematical terms, let L denotes the set of the layers of the ANN given by $L = \{L_1, L_2, \ldots, L_M\}$ where M is the total number of layers. By default, we classify the types of layers into three: input, hidden, and output. Output layer is defined to be the right-most layer which provides the softmax (softmax refers that all individual values, when summed, will equal to unity) probabilities of each class. The number of nodes in the output layer are generally taken to be equal to the number of classes so that each node represents a class. The highest value among all the nodes taken across the output layer provides the predicted class. All the layers falling in between the output and input layers are called as hidden layers which are not observable.

Hence, if there are Mtotal layers, then one may say that there are $M - 2$ hidden layers. We define nodes residing in each layer i by N_i by a set $N_i = \left\{N_i^1, N_i^2, \ldots, N_i^{P_i}\right\}, 1 \le i \le M$, where P_i is the number of nodes in layer L_i . The edges, or links are referred to as weights given by a set $W = \{W_2, W_3, \ldots, W_M\}$. The first element of W is taken to be W_2 because L_1 does not contain any connections, hence, no weights are defined by W_1 . Each set element W_i in W is given by

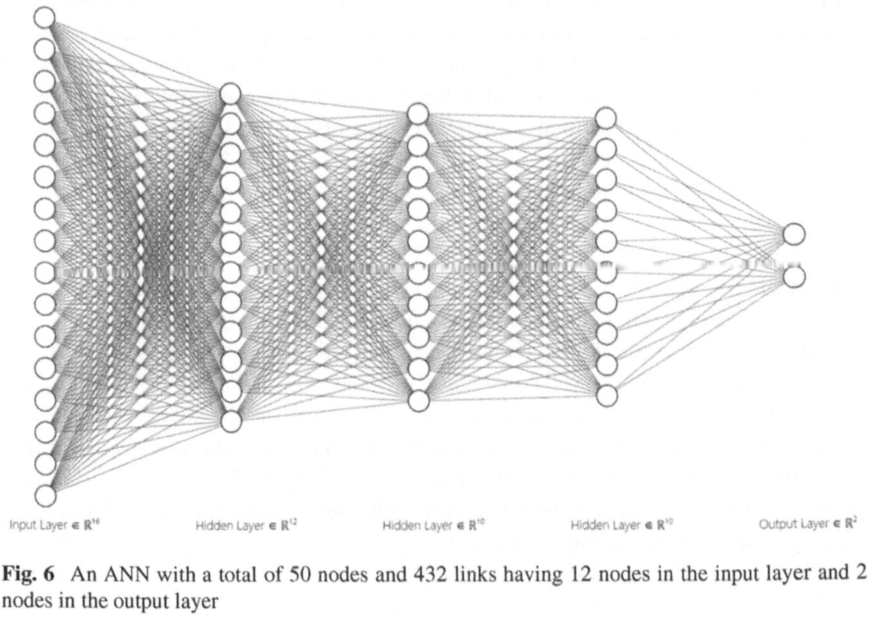

Fig. 6 An ANN with a total of 50 nodes and 432 links having 12 nodes in the input layer and 2 nodes in the output layer

Table 4 Different activation functions

S. No.	Activation function	Equation
1	Sigmoid activation	$\omega(x) = \frac{1}{1+e^{-x}}$
2	Tanh (hyperbolic tangent)	$\omega(x) = \frac{1-e^{-2x}}{1+e^{-2x}}$
3	ReLU (rectified linear unit)	$\omega(x) = \text{argmax}(0, x)$
4	Threshold function	$\omega(x) = \begin{cases} 1 \text{ if } x \geq 0 \\ 0 \text{ if } x < 0 \end{cases}$

$W_i = \left\{W_i^{1,1}, W_i^{1,2}, \ldots, W_i^{1,P_i}, W_i^{2,P_i}, \ldots, W_i^{P_{i-1},P_i}\right\}$. For instance, weight between 4th node of L_5 and 3rd node of L_4 is given by $W_4^{3,4}$. Every node has an activation function embedded through which the decision to either fire or not fire the input signal forward is taken place. The inputs, when passing through nodes and weights, arrive at a particular node and represented by a sum S given by Eq. (6) as,

$$S = \sum_{j=1}^{P_{i-1}} W_i^{j,k} N_{i-1}^j, \ 1 \leq k \leq P_i \tag{6}$$

When inputs are provided at the input layer, these inputs get propagated through the network with random initialized weights, where each node has an activation function $\omega(S)$ helping the node to provide an output. Table 4 describes some of the prominent activation functions used in deep learning.

After a single pass of inputs through the network, a cost function C is evaluated [31] as,

$$C = \frac{1}{2} \sum_b \sum_a \left(\hat{y}_{a,b} - y_{a,b}\right)^2 \tag{7}$$

In Eq. (7), b is an index over cases, or input-output pairs, and a is an index over output units. The network now undergoes a process named backpropagation where the cost C is evaluated and C is minimized. *Gradient descent* is the employed technique in this process, where the partial derivative of C with respect to each weight is calculated. This partial derivative is the sum of partial derivatives of each of the b input-output pairs. The partial derivative of C to the output of the output units is given by,

$$\frac{\partial C}{\partial \hat{y}_a} = \hat{y}_a - y_a \tag{8}$$

Applying the chain rule to ${}^{\partial C}/_{\partial S_a}$ which is the partial derivative of C with respect to the input of the output units representing how a change in the summed input S

affects the error,

$$\frac{\partial C}{\partial S_a} = \frac{\partial C}{\partial \hat{y}_a}\frac{\partial \hat{y}_a}{\partial S_a} = \frac{\partial C}{\partial \hat{y}_a}\hat{y}_a(1 - \hat{y}_a) \tag{9}$$

Now, S is a linear function of weights of the links and states of lower-level nodes. Hence, error variance by a change in weights is given by,

$$\frac{\partial C}{\partial W_i^a} = \frac{\partial C}{\partial S_a}\frac{\partial S_a}{\partial W_i^a} = \frac{\partial C}{\partial S_a}\hat{y}_i \tag{10}$$

The contribution done by the output of the node i to ${}^{\partial C}/_{\partial \hat{y}_i}$ due to the effect of i on a is given by

$$\frac{\partial C}{\partial S_a}\frac{\partial S_a}{\partial \hat{y}_a} = \frac{\partial C}{\partial S_a}W_i^a \tag{11}$$

Summing all the contributions of error from all the connections to unit i, we get,

$$\frac{\partial C}{\partial \hat{y}_i} = \sum_a \frac{\partial C}{\partial S_a}W_i^a \tag{12}$$

After sufficient number of feed-forward and backpropagation steps, the ANN is said to have learnt the complex underlying relationships of the input data to become a very powerful model to regress or classify new data inputs.

B. The Convolutional Neural Network (CNN)

CNNs are one of the most important aspects of deep learning today. CNNs, proposed by LeCun et al. [32], focus mainly on classifying image data based on the spatial pixel-wise relationships of images. This is done through a mathematical concept named as convolution. They comprise mainly two layers—(a) convolutional layers and (b) dense layers. These dense layers actually make up an ANN connected at the output of the convolutional layers. Convolutional layers actually decrease the size of the input dimensions of the images and apply convolution on the data with feature detecting kernels of fixed sizes which are 2D matrices, also known as filters. These obtain feature maps storing the spatial information or relationships between different pixels of the input image which help in classifying the images. Figure 7 describes the structure of a CNN.

The convolution operation of two functions $f(t)$. and $g(t)$ can be defined as,

$$\underbrace{f(t) * g(t) \triangleq \int_{-\infty}^{\infty} f(\tau)g(t-\tau)\mathrm{d}\tau}_{(f*g)(t)} \tag{13}$$

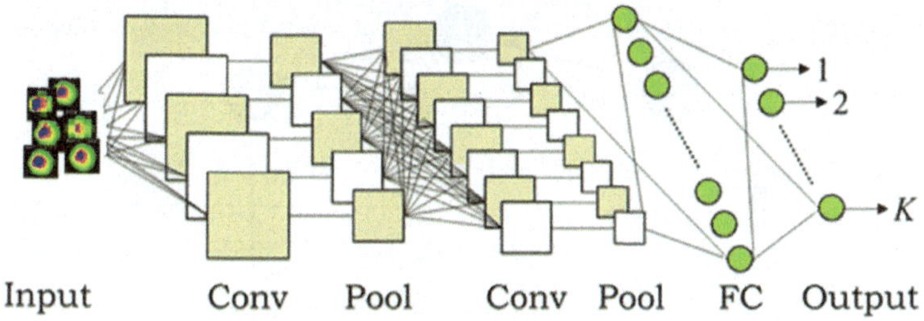

Fig. 7 The structure of a Convolutional Neural Network

This convolution operation is done by the *Conv* layer given in Fig. 6. The input data is downsampled here which is further downsampled by a pooling layer which takes only significant pixel values through a 2D filter. Pooling is done to provide spatial invariance to the CNN model, in other words, it helps classify images even if they are tilted, distorted, flipped, etc. However, it reduces the number of parameters by approximately 75% and also prevents over-fitting of input data. The *FC* term in Fig. 6 refers to *Full Connection* which essentially means that the outputs of the convolutional/pooled layers are flattened and arranged into a vertical array and may even be standardized and normalized to qualify as an appropriate input to the ANN at the back end. CNN has a variety of applications in the era of Industry 4.0. CNN is used different industrial fields as we shall describe at the end of this section, and we see their use in the agricultural industry as well [33, 34].

C. Restricted Boltzmann Machines (RBMs)

In the class of unsupervised deep learning models, restricted Boltzmann machines (BMs) are one of the most significant entities which are also known to be generative models [35]. RBMs are a derivation of the more general Boltzmann machine (BM). BMs are undirected networks, as opposed to ANNs and CNNs. In other words, one may say that there are no input or output layers in BMs. BMs are composed of many nodes linked together through weighted connections. These networks do not have a cost function to be optimized but rather attempt to generate data or probability distribution that closely resemble the input data. There are multiple states generated within the BM and each state is a rough representation of the system that is represented by the data. For our convenience, we define that there are to be three input nodes and seven hidden nodes in the structure of the BM depicted in Fig. 8.

After inputs are fed at the input layer, through contrastive divergence [36] which involves Gibbs' sampling (a Monte Carlo algorithm), the visible nodes iteratively feed the hidden nodes through weights and immediately after which, the hidden nodes feed the visible nodes. One iteration of this cycle is termed as a single Monte Carlo Markov Chain (MCMC) walk since a Markov chain is generated at the visible layer. Practically, creating BMs is computationally very expensive due to the amount of connections required (as each node is connected to every other node) and simply

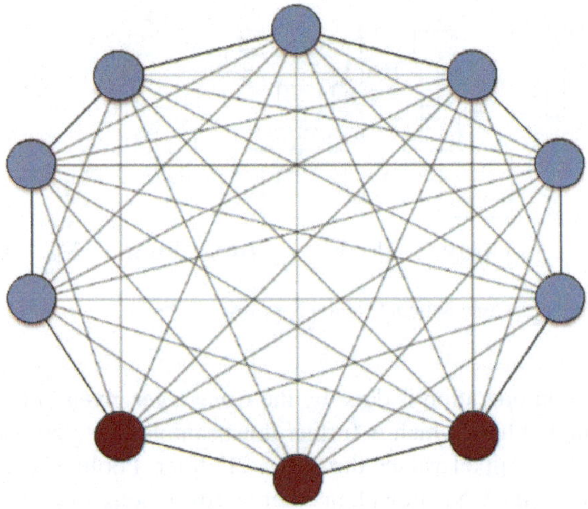

Fig. 8 The Boltzmann machine where blue-gray nodes are hidden and maroon nodes are visible

that with the addition of one node, the number of links increase exponentially. Hence, instead of BMs, RBMs are put into practice.

The change in RBMs that differentiates them from traditional BMs is that the connections between hidden–hidden and visible–visible nodes are forbidden (or 'restricted'). This structure is depicted by Fig. 9.

RBMs and BMs are energy-based models [37] where the energy of the joint configuration of visible and hidden nodes $E(\boldsymbol{v}, \boldsymbol{h})$ is given by [38] as,

$$E(\boldsymbol{v}, \boldsymbol{h}) = -\sum_{i \in \text{visible}} p_i v_i - \sum_{j \in \text{hidden}} q_j h_j - \sum_i \sum_j v_i w_{i,j} h_j \tag{14}$$

where v_i and h_j are the states of the visible node i and hidden node j, p_i, q_j are their biases and weights between them is denoted by $w_{i,j}$. The RBM makes use of the following formula to assign a probability between each hidden and visible vector pair,

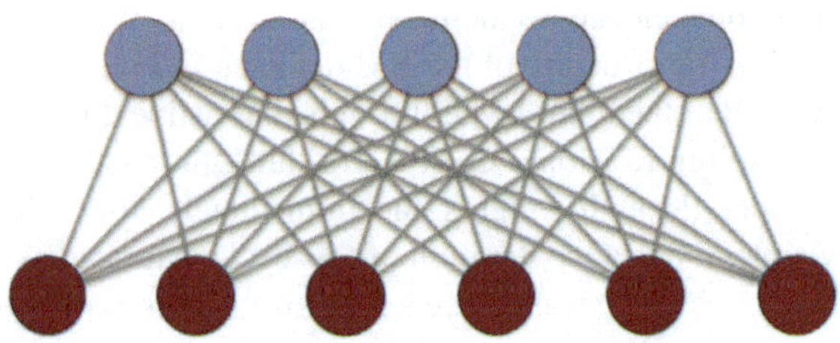

Fig. 9 The Restricted Boltzmann Machine

$$p(\boldsymbol{v}, \boldsymbol{h}) = \frac{e^{-E(\boldsymbol{v},\boldsymbol{h})}}{\sum_{\boldsymbol{v},\boldsymbol{h}} e^{-E(\boldsymbol{v},\boldsymbol{h})}} \tag{15}$$

where, the denominator of the RHS is also called as the partition function. Contrastive divergence follows the gradient for the learning as given by,

$$\frac{\delta \log p\left(v^{\text{initial}}\right)}{\delta w_{i,j}} = \left\langle v_i^{\text{initial}} h_j^{\text{initial}} \right\rangle - \left\langle v_i^{\text{final}} h_j^{\text{final}} \right\rangle \tag{16}$$

D. Autoencoders

Autoencoders have also been classified as unsupervised methods in deep learning but in reality they are *self-supervised*. They also comprise, like ANNs, of three types of layers: input, hidden, and output as shown Fig. 10.

The layers are: input layer, where $\{x^i\}_{i=1}^N \in \boldsymbol{X}$, the middle layer also known as the *bottleneck* or *coding* layer which holds the feature vector $\mathbf{Z}$ and the output layer. Autoencoders are a self-supervised approach to learning lower dimensional feature representations from unlabeled data. The inputs are encoded into feature extracted representations and stored in $\mathbf{Z}$ through weights. Similarly, an output similar to $\boldsymbol{X}$ at $\hat{X}$ is generated after decoding of vector $\mathbf{Z}$. Note that the activation functions of input and hidden layers are the hyperbolic tangent function ($\tanh(x)$). There is a mapping function for encoding which may be stated as,

$$\mathbf{Z} = f(\boldsymbol{W}\boldsymbol{X} + b) \tag{17}$$

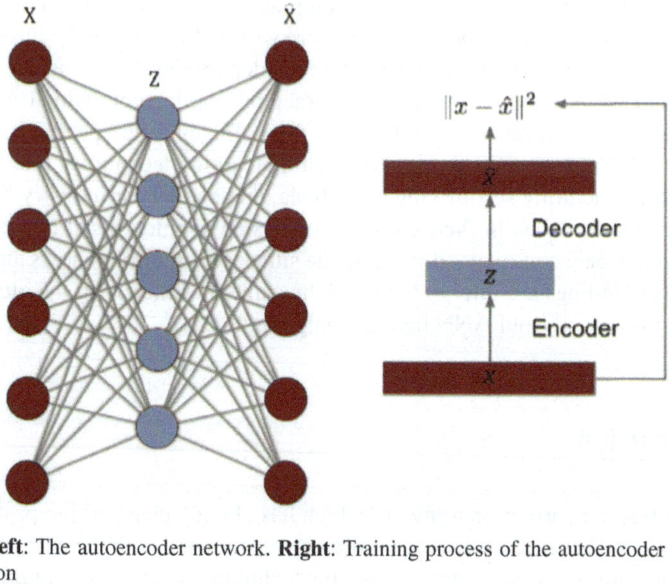

Fig. 10 **Left**: The autoencoder network. **Right**: Training process of the autoencoder with an L2 loss function

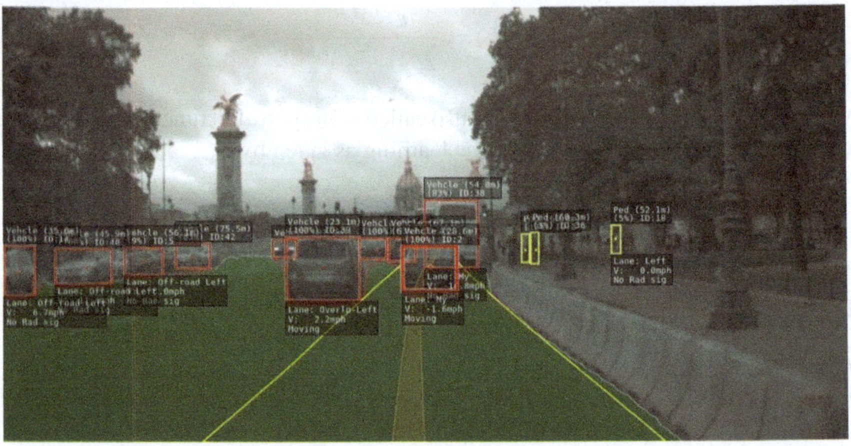

Fig. 11 AI-enabled object recognition for self-driving cars

where b represents bias and vector $\boldsymbol{W}$ represents the weights. The L2 loss function L is defined by Eq. (18) which is calculated at the end of every epoch,

$$L = \left|\left|x - \hat{x}\right|\right|^2 \tag{18}$$

Once the loss is calculated through Eq. (18), like ANNs, the error is backpropagated, and weights are adjusted. Autoencoders are used for compressing data (the data is compressed in the hidden layer) and may be used to initialize supervised classification models where the decoder is replaced with a classifier. This classifier can then be fed the extracted downsized feature vector $\boldsymbol{Z}$ to accurately classify based only on $\boldsymbol{Z}$. Autoencoders can be used in various departments of industry [39–41].

ANN and CNN combined have wide practical applications which are mostly implemented by industries so as to improve their efficiency and products. In recent years, car industries have implemented machine learning techniques like ANN and CNN for manufacturing self-driving cars. Tesla, the leading car industry has implemented these techniques in their cloud-based systems which assist the car how to drive and make decisions depending upon the situation. Figure 11 shows an image of how the surroundings are interpreted from the camera of the car in the cloud-based system and how CNN and ANN help in analyzing it.

5 Conclusion

Industry 4.0 is the current ongoing era which has already changed the past scenario of industries and has made them more efficient, customized with the help of different advanced technologies, and machine learning techniques are the start for the future of many developing industries by creating new business models, cloud-based systems, etc. In this chapter, we have discussed how machine learning has rendered industries more autonomous with various techniques and algorithms. The effect of digitization and advancement in the technology will make living more simple and friendly. The leading developments will create opportunities for the industrial sector to expand rapidly. Digital business models are expanding the existing products and services in order to ensure future growth and development of the industries. In this chapter, concepts related to Industry 4.0 are briefly discussed as well as machine learning techniques and how it can combine with Industry 4.0 to make drastic changes and improvement in the industries to achieve their objectives in the fourth industrial revolution and upcoming revolutions with more advanced techniques. The term Industry 5.0 has been introduced in the research areas which will soon be recognized, as fifth industrial revolution with advanced technology, will include impact on civil society, governance and structure, and a lot more.

References

1. Thangaraj, J., & Narayanan, R. L. (2018). Industry 1.0–4.0: The Evolution of Smart Factories. Retrieved on May 17, 2020. https://www.researchgate.net/publication/330336790_INDUSTRY_10_TO_40_THE_EVOLUTION_OF_SMART_FACTORIES.
2. Nayak, S., Gourisaria, M. K., Pandey, M., & Rautaray, S. S. (2019, May). Recent dimensions of data science: A survey. In 2nd International Conference on Data & Information Sciences ICDIS-2019. (pp. 465–476). https://doi.org/10.1007/978-981-15-0694-9_44.
3. Rojko, A. (2017). Industry 4.0 Concept: Background and Overview. *International Journal of Interactive Mobile Technologies, 11*(5), 77.
4. Bmiet, S. (2018). Industrial revolution—From industry 1.0 to industry 4.0. *Journal of Advances in Computational Intelligence and Communication Technologies, 2*(1).
5. Yin, Y., Stecke, K. E., & Li, D. (2017). The evolution of production systems from Industry 2.0 through Industry 4.0. *International Journal of Production Research, 56*(1–2), 848–861.
6. Xu, L. D., & Duan, L. (2019). Big data for cyber physical systems in industry 4.0: a survey. *Enterprise Information Systems, 13*(2), 148–169.
7. Kamarul Bahrin, M. A., Othman, M. F., Nor Azli, N. H., & Talib, M. F. (2016). Industry 4.0: A review on industrial automation and robotic. *Journal Teknologi, 78*(6–13). https://doi.org/10.11113/jt.v78.9285.
8. Masood, T., & Egger, J. (2019). Augmented reality in support of Industry 4.0—Implementation challenges and success factors. *Robotics and Computer-Integrated Manufacturing, 58,* 181–195.
9. BIS Research, Global augmented reality and mixed reality market: analysis forecast 2018–2025 (2018).
10. He, Z., et al. (2017). Research on human-computer interaction technology of wearable devices such as augmented reality supporting grid work. *Procedia Computer Science, 107,* 170–175.
11. Nayak, S., Gourisaria, M. K., Pandey, M., & Rautaray, S. S. (2020). Comparative Analysis of Heart Disease Classification Algorithms using Big Data Analytical Tool (2020). In *International Conference on Computer Networks and Inventive Communication Technologies (ICCNCT 2019)*, (vol. 44, pp. 582–588). https://doi.org/10.1007/978-3-030-37051-0_65.

12. Nayak, S., Gourisaria, M. K., Pandey, M., & Rautaray, S. S. (2019). Prediction of heart disease by mining frequent items and classification techniques. In *Third International Conference on Intelligent Computing and Control Systems (ICICCS-2019)*. https://doi.org/10.1109/iccs45141.2019.9065805.
13. Nayak, S., Gourisaria, M. K., Pandey, M., & Rautaray, S. S. (2019). Heart Disease Prediction Using Frequent Item Set Mining and Classification Technique. *International Journal of Information Engineering and Electronic Business (IJEEB), 11*(6), 9–15. https://doi.org/10.5815/ijieeb.2019.06.02.
14. Qian, H., Mao, Y., Xiang, W., & Wang, Z. (2010). Recognition of human activities using SVM multi-class classifier. *Pattern Recognition Letters, 31*(2), 100–111.
15. Rejani, Y., & Selvi, S. T. (2009). Early detection of breast cancer using SVM classifier technique. ArXiv preprint arXiv:0912.2314.
16. Dhanalakshmi, P., Palanivel, S., & Ramalingam, V. (2009). Classification of audio signals using SVM and RBFNN. *Expert Systems with Applications, 36*(3), 6069–6075.
17. Jee, H., Lee, K., & Pan, S. (2004, December). Eye and face detection using SVM. In *Proceedings of the 2004 Intelligent Sensors, Sensor Networks and Information Processing Conference, 2004.* (pp. 577–580). IEEE.
18. Chen, K. Y., Chen, L. S., Chen, M. C., & Lee, C. L. (2011). Using SVM based method for equipment fault detection in a thermal power plant. *Computers in Industry, 62*(1), 42–50.
19. Pal, M. (2005). Random forest classifier for remote sensing classification. *International Journal of Remote Sensing, 26*(1), 217–222.
20. Bernard, S., Adam, S., & Heutte, L. (2007, September). Using random forests for handwritten digit recognition. In *Ninth International Conference on Document Analysis and Recognition (ICDAR 2007)* (Vol. 2, pp. 1043–1047). IEEE.
21. Patel, R. K., & Giri, V. K. (2016). Feature selection and classification of mechanical fault of an induction motor using random forest classifier. *Perspectives in Science, 8,* 334–337.
22. Abdullah, A. S., & Rajalaxmi, R. (2012, April). A data mining model for predicting the coronary heart disease using random forest classifier. In *International Conference in Recent Trends in Computational Methods, Communication and Controls* (pp. 22–25).
23. Erman, J., Arlitt, M., & Mahanti, A. (2006, September). Traffic classification using clustering algorithms. In *Proceedings of the 2006 SIGCOMM Workshop on Mining Network Data* (pp. 281–286).
24. Yeung, M. M., Yeo, B. L., Wolf, W. H., & Liu, B. (1995, March). Video browsing using clustering and scene transitions on compressed sequences. *Multimedia Computing and Networking, 2417,* 399–413.
25. Dhanachandra, N., Manglem, K., & Chanu, Y. J. (2015). Image segmentation using K-means clustering algorithm and subtractive clustering algorithm. *Procedia Computer Science, 54,* 764–771.
26. Ng, H. P., Ong, S. H., Foong, K. W. C., Goh, P. S., & Nowinski, W. L. (2006, March). Medical image segmentation using k-means clustering and improved watershed algorithm. In *2006 IEEE Southwest Symposium on Image Analysis and Interpretation* (pp. 61–65). IEEE.
27. Elango, M., Nachiappan, S., & Tiwari, M. K. (2011). Balancing task allocation in multi-robot systems using K-means clustering and auction based mechanisms. *Expert Systems with Applications, 38*(6), 6486–6491.
28. Maqbool, O., & Babri, H. (2007). Hierarchical clustering for software architecture recovery. *IEEE Transactions on Software Engineering, 33*(11), 759–780.
29. Gourisaria, M. K., Das, S., Sharma, R., Rautaray, S. S., & Pandey, M. (2020). A Deep Learning Model for Malaria Disease Detection and Analysis using Deep Convolutional Neural Networks. *International Journal of Emerging Technologies, 11*(2), 699–704.
30. Rosenblatt, F. F. (1958). The perceptron: a probabilistic model for information storage and organization in the brain. *Psychological Review, 65*(6), 386–408.
31. Rumelhart, D. E., Hinton, G. E., & Williams, R. J. (1986). Learning representations by back-propagating errors. *Nature, 323*(6088), 533–536.

32. LeCun, Y., Bottou, L., Bengio, Y., & Haffner, P. (1998). Gradient-based learning applied to document recognition. *Proceedings of the IEEE, 86,* 2278–2324.
33. Sharma, R., Das, S., Gourisaria, M. K., Routray, S. S., & Pandey, M. (2019). A model for prediction of paddy crop disease using CNN. In *2nd International Conferences on Computing analytics and Networking (ICCAN-2019)* (pp. 533–543). https://doi.org/10.1007/978-981-15-2414-1_54
34. Rautaray, S. S., Pandey, M., Gourisaria, M. K., Sharma, R., & Das, S. (2020, March). Paddy Crop disease prediction—A Transfer Learning Technique. *International Journal of Recent Technology and Engineering, 8*(6). https://doi.org/10.35940/ijrte.f7782.038620
35. GM, H., Gourisaria, M. K., Pandey, M., & Rautaray, S. S. (2020). A comprehensive survey and analysis of generative models in machine learning. *Computer Science Review, 38,* 100285. https://doi.org/10.1016/j.cosrev.2020.100285.
36. Woodford, O. (2006). Notes on Contrastive Divergence. Retrieved from http://www.robots.ox.ac.uk/~ojw/files/NotesOnCD.pdf. Accessed 19 May 2020.
37. LeCun, Y. (2006). A Tutorial on Energy-Based Learning. Retrieved from http://yann.lecun.com/exdb/publis/pdf/lecun-06.pdf. Accessed 19 May 2020.
38. Hopfield, J. J. (1982). Neural networks and physical systems with emergent collective computational abilities. *Proceedings of the National Academy of Sciences, 79*(8), 2554–2558. https://doi.org/10.1073/pnas.79.8.2554.
39. Zhu, Z., Wang, X., Bai, S., Yao, C., & Bai, X. (2016). Deep learning representation using autoencoder for 3D shape retrieval. *Neurocomputing, 204,* 41–50.
40. Deng, L., Seltzer, M. L., Yu, D., Acero, A., Mohamed, A. R., & Hinton, G. (2010). Binary coding of speech spectrograms using a deep auto-encoder. In *Eleventh Annual Conference of the International Speech Communication Association.*
41. Lu, C., Wang, Z. Y., Qin, W. L., & Ma, J. (2017). Fault diagnosis of rotary machinery components using a stacked denoising autoencoder-based health state identification. *Signal Processing, 130,* 377–388.

Mahendra Kumar Gourisaria is presently working as an assistant professor in the School of Computer Engineering at KIIT University, Bhubaneswar, Odisha. He has received his master degree in Computer Application from Indira Gandhi National Open University, New Delhi, and M.Tech in Computer Science and Engineering from Biju Patnaik University of Technology, Rourkela. He is pursuing his Ph.D. from KIIT Deemed to be University. He has an experience of more than 18 years in academia and 7 years in research. He has published more than 25 research papers in different international journals and conferences of repute. He has also served as the organizing committees members of various conferences and workshops. His area of research includes cloud computing, data mining, soft computing, and Internet and web technology. He is a member of IAENG, UACEE, and life member of ISTE, CSI, and ISCA. He can be reached at mkgourisaria2010@gmail.com and mkgourisariafcs@kiit.ac.in.

Rakshit Agrawal is currently pursuing B.Tech in KIIT Deemed to be University and is currently in the first year. He has won gold medal in robotics and various other state-level robotics competitions. His research interest includes machine learning, neural networks, deep learning, and cloud computing.

Harshvardhan GM is currently pursuing B.Tech in KIIT Deemed to be University and is currently in the second year. He is adept at joint collaboration with various other researchers in the field of machine learning and cloud computing. He is a published author with his survey paper on generative models in Computer Science Review journal with many other works in their final stages or conditionally accepted for publication. His research interests include unsupervised and self-supervised deep learning methods, clinical applications of machine learning for medicine, generative models in machine learning, recommender systems for content delivery, energy-efficient task scheduling in cloud computing, topic modeling in natural language processing using latent Dirichlet allocation, and Internet of drones. He has won various national and state level awards for literary skills involving essay competitions and vocabulary quizzes. He can be reached at harrshvardhan@gmail.com.

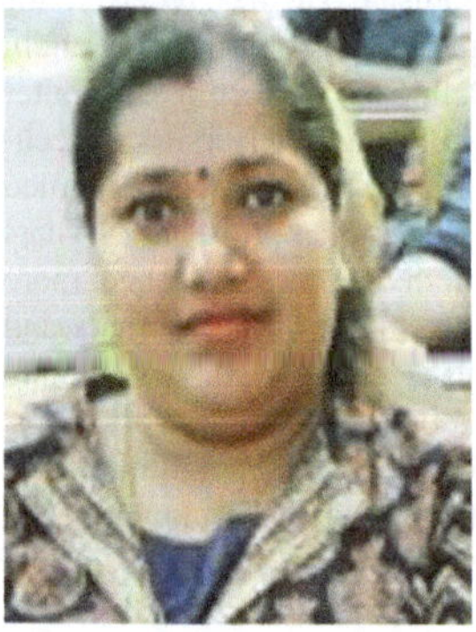

Manjusha Pandey is presently working as an associate professor at School of Computer Engineering, Kalinga Institute of Industrial Technology (KIIT), Deemed to be University, Bhubaneswar Odisha. She did her doctoral degree in Information Technology with MHRD scholarship from Indian Institute of Information Technology (IIITA), Allahabad, U.P. She has an experience of more than 9 years in academia and research. Her research interest includes big data analytics, wireless sensor networks, data mining & warehousing, and similar innovative areas. She has received the research grant from ICSSR under IMPRESS scheme for the project title: Intervention of Use of Technology in Government Sponsored ICDS: A case study of Khorda District as Project Coordinator. She has published more than 80 research articles in reputed journals, conferences, and book chapters. She has delivered 6 invited talks during FDP and in other orientation programs. She has organized a number international conferences, workshops, and invited talks. Presently, she is the faculty coordinator of the data science group at School of Computer Engineering.

Siddharth Swarup Rautaray is presently working as an associate professor at School of Computer Engineering, Kalinga Institute of Industrial Technology (KIIT), Deemed to be University, Bhubaneswar Odisha. He did his doctoral degree in Information Technology with MHRD scholarship from Indian Institute of Information Technology (IIITA), Allahabad, U.P. He has an experience of more than 9 years in academia and research. His research interest includes big data analytics, human-computer interaction, data mining & warehousing, and similar innovative areas. He has published more than 85 research articles in reputed journals, conferences, and book chapters. He has delivered 5 invited talks during FDP and in other orientation programs. His research citation index is more than 1600. He has organized a number international conferences, workshops, and invited talks. Presently, he is the faculty coordinator of the data science group at School of Computer Engineering.

Web Semantics and Knowledge Graph

Bharat Sharma

Abstract In this modern era of big data, according to the researchers, it is estimated that almost 80% of the enterprise data is unstructured data, which means it is in the form of documents, research reports, surveys, articles, slides and even on the emails, etc. In the crucial time our world is where technology reign supreme, data is considered as the most important resources and even though we are surrounded by data, we cannot analyze it because of our traditional data technologies. Although there are new technologies on the way yet that is far from technologies, we need to unleash the full potential of data. Most of the unstructured data which comes from documents, reports, etc. are in the form of language phase, and it can entirely be a natural language sentence. In order for it to be analyzed, we need to transform, extract and process the relevant information from it. Hence for searching or merging the data from different data resources or platforms, semantic technology comes into the picture. Semantics application is steered by knowledge graph and that you do not need to go to hassle of data migration, as semantics connects it. Semantics can be used for all standard-based technologies removing the need for using different technologies for different data forms. While there are different enterprise catalogue products in the market, there is not a solution that fully understands the context of a user's input, it can be either from a domain perspective, the depth of the knowledge the user is asking for or just comprehensively addressing the questions coming from the user perspective. For visualization purpose, you can compare semantic technology to a Google search; it is comparable in terms of results though Google uses natural language understanding (NLU), natural language processing (NLP) and other rich metadata. Semantic technologies also use some of the key points like metadata enrichment, ontologies and search and graph technologies to achieve the same feat and understanding the question from user perspective. For specific domains, there can be EKG or enterprise knowledge graph which consists of data which can be structured or unstructured. It can use information repositories and ontologies for building better domain-specific graphs. The end goal of semantic technologies is to make machine understand the data; semantic technologies use widely accepted RDF. Knowledge

B. Sharma (✉)
ZS Associates, Pune, India
e-mail: yashsharmamgs@gmail.com

M. Pandey and S. S. Rautaray (eds.), *Machine Learning: Theoretical Foundations and Practical Applications*, Studies in Big Data 87,
https://doi.org/10.1007/978-981-33-6518-6_5

graphs are specific graphs containing the nodes and edges containing the useful information for a particular domain. Subject matter experts can be consulted with while building the knowledge graph in order for it to contain all the domain information it needs for answering the queries modeling knowledge domains can be considered as the core for semantic technologies, knowledge models defines entities, relations, attributes and values. Semantic technologies can also be integrated with machine learning models for in order to increase the effectiveness of the search engine. RDF or any other data model can be processed in a way of how human thinking works making it effective as it contains triples, i.e., subject, predicate, object which can be just natural language statements or very close to natural language phase.

Keywords Semantics · Search · Taxonomies · Ontologies · Search engines

1 Introduction

While working in the corporate sector, the questions that always come to our mind are: Why to embrace semantic technologies? And how to implement semantics solutions? For answering that firstly, we will see the challenges with enterprise architecture.

Data Silos: Data silos can be explained as the collection of information that is isolated from other parts of the organization. Data silos stops the user to get the comprehensive overview of the data and results in wasted resource and inconsistent data. *Limit-real-time data analytics functionalities*: Whenever there is a real-time data analytics, the architecture limits it. There can be many reasons behind the problem, but the result is always an organization limit is capability of analyzing data.

New queries based on relational database require resource: Every time a new query is fired onto the database to get some resource, it mostly results in consuming some resource out of it. *Cross-function effort is time consuming*: Even if we try to remove all the limits, the efforts it takes are really time consuming.

Different schemas complicate data migration projects: All the different schemas used while building the database makes it harder to integrate it under a single umbrella and makes it complicated even if there is a need to fire some of the less complicated of queries.

The problems described above are not all and there are a lot of hurdles in the enterprise architecture but the majority of the problems lies under these five major issues. Now it is simple that the architecture used in the corporate sector is far from efficient and this is when semantic technologies come into the picture and save the day (1.1).

1.1 Advantages of Using the Semantic Technologies

We know the hurdles in the enterprise architecture, and now let us discuss how semantic technologies can be used to remedy those and all the advantages that the semantics technologies can offer.

Standard-Based Technologies: Semantic technologies are through the standards set by the World Wide Web Consortium (W3C). *No vendor lock-in*: As with the vendors when the technologies are being used, there is supposed to be lock-in period in which you cannot use other technologies.

Transforming structured and unstructured data with RDF: Semantics technologies enable meaning to be attached to data; either it be structured data or unstructured data it can be integrated into the semantic technologies without the use of the migrating the entire schemas. It gives all freedom which data to combine regardless of the state of the data.

Sophisticated queries due to graph structure: As semantic technologies are often steered by the knowledge graph, it uses the advantage of the graph structure, so even the complex queries give a better result. *Subject matter experts included in data-related tasks*: While building a data model or an ontology, SMEs or subject matter experts are called upon so that their expertise can be used to model the data, while building ontologies also, the SME are called.

Incremental approach to develop a linked data environment: With linked data, we can build long chains of build network of statements which usually are very similar to the natural language, and we have leading to incrementally updating the network and increasing the range of questions or queries it can answer to.

1.2 Challenges in Note

Semantic technologies are not used to substitute the existing technologies because it was introduced for a specific use case; there can be scenarios where semantic technologies can be used and the performance might not increase significantly or it might not be beneficial.

Data models can be processed in a way that how the human thinking works. How humans process thoughts and answer to certain things. If we combine the power of the knowledge modeling and text analytics, it can be used as the foundation for machine learning and human interaction. Now coming to the basic idea of knowledge graph, we use knowledge graph in day-to day-life without even knowing, for example, the knowledge graph used by Google is called Google info graphs, LinkedIn has economic graphs which matches data with skill professional experts, Facebook has its social graph containing entities such as friends and family. Knowledge models can define entities with the relation and its attributes as the values. While building a knowledge model, we must keep this in mind that it is not necessary to start with a very complex data model; actually, it is quite the contrary we can start small and while building, the knowledge graph can increase the complexity of the data model step by step.

2 Basics of Taxonomies

In the above sections, we have discussed the usage of web semantics and knowledge graphs, but the basics of knowledge graphs are not yet explained; this section gives you a brief idea about taxonomy and what factors are necessary while building the knowledge models.

Folksonomy can be defined as the set of keywords that are assigned to an information source by the crowd itself; collaborative tagging is the process of tagging itself. Folksonomy [1] is not just metadata about the content, it complements information source itself. It highlights more aspect about the information source, maybe other meaning associated to it. Interesting aspect to folksonomy is that since it is governed by the crowd, it is mostly constantly changing. Folksonomy allows user to add tags in the content management system which is quite useful for our model but sometimes, it can be totally unrelated to our model. It can be a useful information but just not for our model. Controlled vocabulary is also used for tagging content in the content management system, it is steered to be useful specifically for the knowledge model, and it can be glossary in matter of knowledge share or can be a meaning related to it or an information (Fig. 1).

A taxonomy [2] can be set of definitions that are organized in a hierarchy, it starts with the general description of something, and it gets more defined and specific you go down in that order. We could see the categories, it models the broader concept, and after each level, the information gets precise. It is a machine-readable model of

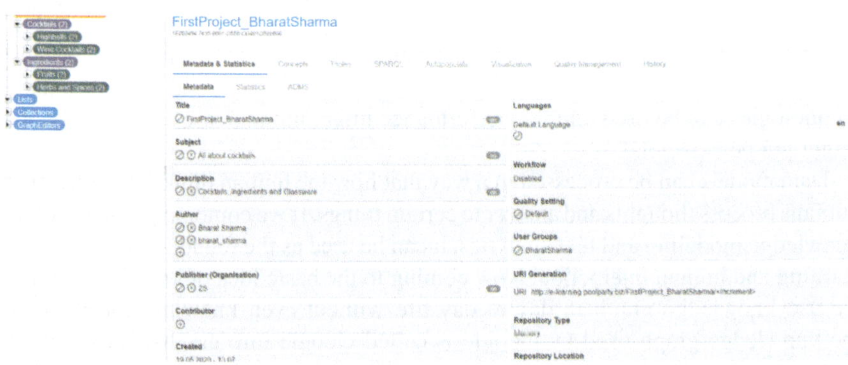

Fig. 1 Thesaurus PoolParty

Fig. 2 Animal kingdom

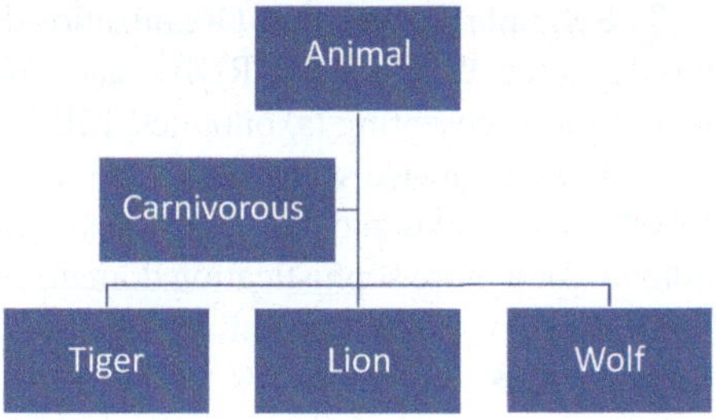

a knowledge domain because it uses RDF and URI which is explained in the next section. It can be model of a small part of the world, for example (Fig. 2).

Taxonomy can describe things that are important for a domain, for example, if the domain is animal, the taxonomies can be like the example above. Taxonomy can be used to describe the relationships between those things. Taxonomies can be base for the knowledge model and applications can be built on top of it.

Knowledge models contain structural information about the domain, and domain could be anything so that you are building your model for it. Knowledge models can be simple knowledge model where there is flat list of terms that say something about the domain, but it does not exactly define what it is, it can be compared to the metadata, and it has useful information. On the other hand, taxonomy is the more expressive knowledge models. It clearly defines all relationship between the concepts; being in the hierarchical structure also enables it to focus on the broader domain. For example, in the given above example if the question being, whether tiger is an animal? It can be easily looked upon and answered as tiger is a carnivores animal. It can be something different also, for instance, name one carnivore's animal and the answer can be easily deduced to tiger. Being in the hierarchical structures gives taxonomy another benefit that it grows according to the needs. For instance, if you want to add another animal in the list of carnivores, or in some case if you want to increase the complexity of the model which means you want to define custom relationship with the concepts in that case, all you need to do is define some custom relationship.

2.1 Advantages of Using a Taxonomy

As we discussed the basics of the taxonomy, now it is the time to see the advantages of using taxonomy.

To start with all the concepts are embedded in a context that means when you see that tiger is in the list, you can see the concept and understand its context. Also, information can be aggregated easily on the given concept and also information can be added onto any level being in hierarchical order, it does not disturb the concept of others.

The Simple Knowledge Organization System (SKOS) is a W3C standard that builds on the W3C's RDF, RDFS, and OWL specifications. To provide a standard model for representing taxonomies, FIBO ontologies are based on the SKOS which specializes in finance sector, and RDF schema are base for the SKOS schemas. With SKOS, taxonomies and classification systems are made, when you are narrowing the domain for a more sophistication domain for the schemes and ontologies. Then after you can use the underline ontologies and schema for the purpose of text mining and data analytics.

2.2 *Components of the Taxonomy*

Concepts: It can be a word, a term maybe a label, it is a concept which is imbedded in the broader sense, concepts itself can be a synonym, maybe labels, or any kind of language tags, precise information cannot be deduced in the form of language tags, for example, Amazon, it can be anything from the concept perspective.

Concept schemes: These are basically the broader domain maybe categories, giving the concept some information and meaning. *Top concept*: They are also concept, but they can be used to define which domain the concept schemes is expanding into; concepts basically define the top concepts, for instance, in our below case (Fig. 3).

Fig. 3 Amazon forest scheme

- ❖ Amazon (Concept Schemes)
  - ➢ Forest (Top Concepts)
    - ▪ Animals (Concepts)
    - ▪ Trees
    - ▪ Rivers
  - ➢ Living Area
- ❖ Amazon (Concept Schemes)
  - ➢ Animals (Top Concepts)
    - ▪ Lions
    - ▪ Tiger
  - ➢ Rivers
    - ▪ Nile

3 RDF Data Model

These are the basics for all the semantic web technologies, RDF or resource description framework is close to natural language statement because RDF [3] uses triples, and it consists of subject, predicate, and object which are framed in such a way that it resembles the natural language and it shows how information can be extracted from it.

Example for a triple can be anything like, Rohan likes Cricket. In the given example, Rohan is the subject for which the information is given, predicate is the relationship between the subject and the object (Fig. 4).

Natural language statement can be broken down into triples and then can be used. With linked data, long chains networks of statements can be built up accordingly and natural language can be formatted in that way. RDF data model is built on the idea of building resources on its very core. While querying the data, these predicates play a vital role, data can be querying based on the relationship.

URI or uniform resource identifier is the link between the resources that can be used to uniquely identify from each other. By resource, it meant subject, object, or predicate; all the things are considered as URI. It is because of these URI that the RDF data model is machine readable. URI is assigned to each predicate, URI is a link or can be an ID, and it needs to be unique in order to be used. Best practices while assigning a URI are mostly to use resolvable URL as they can be used to receive information and they are distinguishable from each other; while reading the URL, the resource information can also be seen. For our above example Rohan will have a URI, Likes will have a URI and Cricket will also have a unique URI, for any subjects best practice involve adding an alternate label or a URI.

RDF also defines some elements that can be used to describe resources, for instance, rdf:type for Rohit can be human or a person. These are used to answer the questions that is Rohit human?

RDF ontology can be a very basic ontology; RDFS or resource description framework is the extension for the RDF ontology.

RDF serialization formats are standard for how RDF data can be written down or stored, written down and exchanged, it is just like the format in which you want the data to be stored, and above explained triples are just another common form to store the data. Some other formats for storing the data are (running).

Triples: As explained above, it is the most common form to store the data, and it consists of subject predicate and object. *N-Triples*: These are quite similar to the triples with the exception that they are easy to parse line-base format and they are marked with double quotes and end with a ".".

Fig. 4 Subject–predicate relationship

N quads: It is an extension to N-triples and used for serializing multiple graphs. *JSON-LD*: These are JSON format based for storing the data.

XML: These are the XML-based format, and initially, the only RDF format available was XML.

4 Basics of Ontologies

An ontology can be defined as the formal description of knowledge. It has specific relations and attributes, which enables it to link multiple concepts, and it has highly personalized content and can be used to find results for a dynamic question with using multiple concepts. An ontology is a formal naming and definition of the types of properties and the interrelationships of entities in a domain. Ontology [4] has the basic three components; it defines the types and defines the properties and the relationships between them.

For type, it can be the classes which can be considered as the concept or maybe a set of concepts which share properties and relationships, it can also have subcategory in classes like subclasses, and the point to emphasize on is that the exact or the actual value is not defined in the ontology. Example for this can be Cricket teams or teams for that say, the classes have a broader domain and the properties define the attributes that have information of the concept, it can describe concept or can have additional information about the concept. After classes and attributes next key element comes the relationships which work as a link between them, relationships precisely specify the kind of relationship class have with the attributes. Relationship gives context between the classes and attributes like Cricket is the class in the above example and attribute is providing some information about the class, i.e., Cricket has 11 players, it is the "has" relationship which provide the context for the class and the attributes if the relationships change, the context varies accordingly, Cricket does not have 12 players, i.e., the context of the said example links directly to the relationship making reader aware the class has this attributes. Below are some examples to explain the topic.

Cricket (Can be class) or Cricket Team (Class).

For attributes it can have number of players, i.e., Cricket has 11 players.

Relationship provides the context of the class with the attributes, i.e., Cricket (Class) "Has" (Relationship) 11 players (Attributes) (Fig. 5).

SKOS ontology: It is described as the most basic ontology, it can define the properties each resource can have, tools like PoolParty have a visual dashboard where you can see the type of relationship that class has, it helps user appeal and check for a class whether the required attributes are there or do they need to add more attributes, is the relationship between the class and the attributes is clear or do they need to have some more definitions for the class.

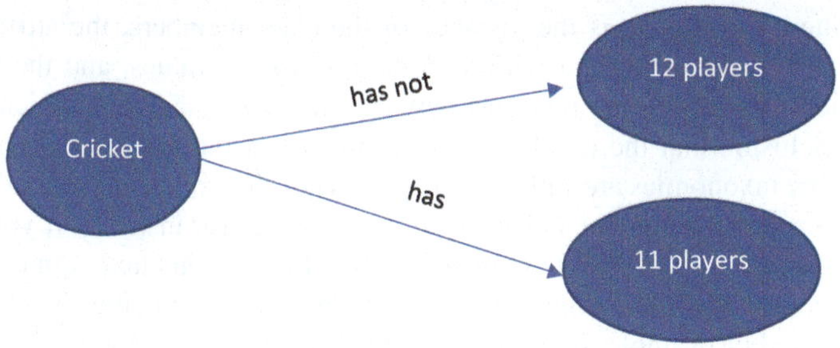

Fig. 5 Has relationship

4.1 Classes and Relationships

Classes can have the skos:Concept, which defines concept as part of a class and can also be used to understand that the concept is subpart of the class.

Attributes can have skos:prefLabel, i.e., these are the labels the attributes are generally called with, or the attributes can have the skos:altLabel, i.e., the alternate labels the attributes are called with. For a layman perspective, these can be the preferred name a person likes someone to call them with, like "IP" for Ishan p, and on the other hand, alternative labels can be like "ishu" for Ishan. It helps identifying the different labels the attributes can have.

Relationships show the domain in which they are going in, for example, skos:narrower or skos:broader, for showing that whether the relationships are getting narrower or broader. For example, grapes is a fruit then for the attribute grapes the relationships is getting broader in the domain of fruit on the contrary, grapes are green denotes the narrower relationships. Another relationship can be skos:related or skos:topConcept of Which are self-explanatory after the above context.

The thing to keep in mind is that different classes have different relations with the attributes. For instance, grapes have sweetness or Rohan is sweet, which also implies in simpler terms that you can also restrict certain attributes to certain class with the help of relations and it will be used for that context only. To not confuse things, it simply means that with different relations, if a class has certain attributes, it is not necessary that the attributes need to pass to another class in the same mode. It does not mean that the relationship will restrict it to other models also.

5 Taxonomy and Ontologies, the Difference and Similarity

Now it is easy to get parallelism between the ontology and the taxonomy as the ontology has the class, attributes, and the relations, whereas the taxonomy stores everything in the form of triples, i.e., subject, predicate, and object. This is how taxonomy [5] and ontology work hand in hand; the classes can be explained as subject where the subject is the instance of the class members, the attributes can be explained as the object but with a specific attributes values, and the predicate just to underline can be the relations between the class members, which draws a simple conclusion that the ontologies are a subpart of the taxonomy or in simple word specific taxonomies are called ontologies. They do work hand in hand but their usage can differ based on the situation they are used as, for instance, if you want to improve search, i.e., personalize searches, then ontologies and taxonomies are to be worked together. Models change and it needs to be adapted with the change with the business case. For example, if you start establishing the search application [6], then taxonomy will suffice or if you want to start small, then based on situation sustainable merge them onto something other basic search engine then start with taxonomy and then you can narrow it down to the ontologies you need.

With the help of custom schemes and ontologies, taxonomy can be extended upon. For instance, if you want to extend the related concept so that the taxonomies can have a broader concept, or if you introduce related concept with the ontologies, for example, if we have a Cricket ontology which we describe about the Indian team and the players it can be added on with the taxonomy of Cricket and it will increase the domain of the taxonomy. It can also define specific relations, for instance, if the taxonomy Cricket has the related column in that case, the ontology will just add Indian team and number of times it has won a cup.

It can also be described with broader and narrower schemes, for instance, if we just add up the Indian team and the taxonomy was about the rules and governing body, then it has the skos:broader relation with team and now the taxonomy Cricket has another body to add into the top concepts with teams, in which the ontology Indian team provides details about the team, members, and so on. Or we can just add the narrower concept of particular team players and the skos:narrower can be used to narrow down the taxonomies to a particular player in action. Also custom schemes can be used to create a mashup of ontologies; this helps with expanding the domain for the taxonomy, for instance, the said ontology Cricket can have multiple set of ontologies or a mashup of ontology in which the teams current player, their record or maybe the list of retired players, another ontologies can have their basic stats about how many matches they have played and so on. The mashups of the above ontologies will create a taxonomy that has far more information and details and can be used to create a knowledge graph where you can see the stats of the players all of their stats, their records, and the list of the retired players of the specific domain, the teams can be the top concept, and they can have attributes like players where as another top concept can be rules where all the attributes revolve around batting, fielding, and bowling. With the mashup ontologies, another advantage is that it is not necessary to add up all the details, the details which are relevant to the taxonomy can only be used. When building a taxonomy, the central component or the base can be ontology which is made centrally, and the taxonomy is built on top to give more relations and extend attributes and attributes types. It is best practice to build a custom schema on top of the ontology before using it for building a taxonomy; custom schemes refine the ontologies for the taxonomy and hence improve the domain information for taxonomy. Importing an ontology is quite easy in tools like PoolParty either you can use system-based ontologies, or you can download the ontologies from free sources, or you can handmade your ontology which suits your taxonomy. Ontologies can be made using any approach but to face less complexity, it is either built up in the top-down approach where concepts are defined and then the attributes are added, or it can be the top-down approach. For the schemas and ontologies, Schema.org is a website which can be referred from and it is maintained by Google so there are very vast collections you can find it there. For instance, there are tailor-made ontologies for use, like the geolocations there are ontologies which can be directly imported to increase the domain knowledge of taxonomy or creating a specific knowledge model on the above.

6 Introduction to Text Mining

In the previous sections, we covered all the basics about the ontologies, taxonomies, how one can create a taxonomy and ontology and use it and more. Now, we will implement it into the field of text mining and see how text mining works. But before that, we will be giving some basic linguistic concepts for text mining.

Linguistics can be strictly described as the study of language, natural languages in which human communicates and expresses themselves. Talking about language, the basic unit of languages is word which is normally referred as "term" in linguistic world; now, these words also have different forms which are used from time to time depending on the usage and the emotion or meaning it is conveying, not every word is of the same form, word forms are all inflection of a word that are generally used or are possible in a language to be used to have a proper conversation. Lemmatization is the process of identifying and grouping of different word forms under a common context or form. For example, if you group different form of create, it can be creation, it can be creative, it can be creatively, if we group them (create, creation, creative, creatively), this is lemmatization. But when working with the natural language even if we group word forms, there is still a bigger challenge remaining, i.e., challenge of ambiguity, this is nothing else then the context in which word is used for can carry different meaning then the word is usually used in the context, for instance the word "break" can have multiple meaning depending upon the context in which it is used, it can be an abrupt occurrence that is interrupting something, or it can be a the cracks of the earth crust, it can be the order to destroy something. Other challenges are homograph; they are words that are spelled the same, but they have entirely different meaning when used with a different context. For example, with keeping year 2020 in mind, "bat" it can be bat used in Cricket (a part of sports equipment) or bat, mammals of order Chiroptera, vampire like animals. Homophones, are the words that are pronounced the same but have different meaning for example cell or sell. Homonyms are words that are spelled and even pronounced the same, but they have different meaning based on the context in which they are used band; it can be the musical band or the band which a person wears on hand or head. Synonyms are another concept; they are different words with similar or close to similar meanings and can act as a replacement for the word they are used for. Synonyms are useful in case of text mining as they cover different words, and hence, they are useful to improve the performance and sometimes also increase the extraction results in text mining.

N-grams are sequence of words that have a joint meaning, i.e., if these words are used independently, they all have different meaning but when they are used in the combination; they have a joint meaning like bigram which are sequence of two words, trigram are sequence of three words. There are tons of example online; they are extensively used when developing a knowledge model for text mining, like "medium blog" a bigram, or "medium blog post" is a trigram. N-grams help in resolving the ambiguity in the sentence meaning and help in increasing the performance while extracting meaning. Co-occurrence is the frequency with which a term occurs together in a text, in comparison to the frequency with which they occur in their own in the same context. These are pretty helpful when identifying the n-grams in the context and discover all. Co-occurrence is helpful in analysis also to check whether the context has any ambiguity or not. Stop words are the words that have a linguistic functions but they generally do not carry any context or meaning of their own; generally, they appear between words to link the sentence and are frequently used in the context, for example, being of, to, by, or, as, for these are used in a sentence but it does not hold value for the user. For a text analysis, they are identified and are sometimes removed so that the context of the sentence can be understood.

Text corpus is the collection of texts about one topic, you can say text corpus as groups of texts carrying information or context about one topic. Corpus analysis is the analysis of text corpora or text corpus. Corpus analysis is another handy technique which can be used to reliably identify n-grams and understand the context of the same; they look at the co-occurrence of and analyze with respect to a large number of texts to see whether there are n-grams in the collection of the text about the topic. Corpus analysis can have multiple uses like it can be used to group different word forms or it can be used to check any specific key terms that are important for the topic. Corpus analysis is useful in extracting in the context of the given text. Reference corpus can be used to extract terms which will make the candidate concept; now if the candidate concepts are seem viable for the consumption into the taxonomy, it will enrich the taxonomy with skos:concept. Corpus analysis can be used for detecting new evolving vocabulary; it can be used to check for the trending topic which can be used to improve the domain and suggestion for the taxonomy. Creating a web corpus is easy it is like the database which has all your information for the topic you are creating it for. Corpus extraction can be done using different methods, you can add document for which you want to extract from. In tools like PoolParty, you can see a dedicated tools where you can extract your text corpus, it has the corpus management system, you can have documents uploaded and you can do the corpus analysis it will add the corpus summary, which will contain all the list of words it has extracted from and all the concept it has. For corpus analysis, it depends on how much depth you want your analysis to be, it can be a few mb files to hundreds of mb depending on the requirement of the analysis. It more depends on how varied your documents are based on the topic you want your analysis to revolve around,

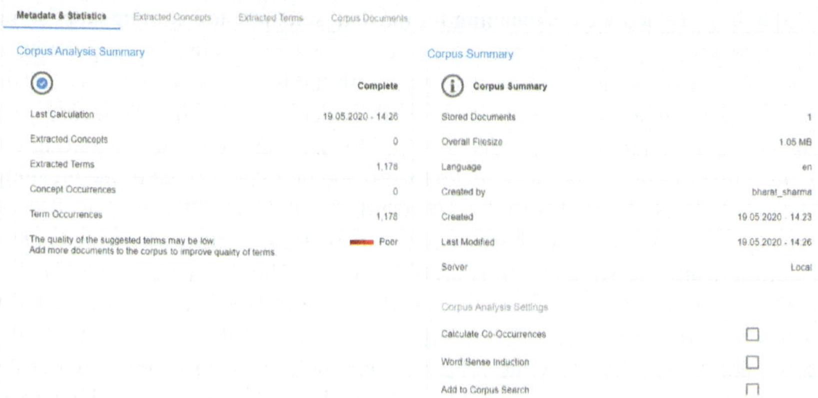

Fig. 6 Corpus analysis

it should be related to the topic else it will reduce the corpus analysis. When the corpus analysis is completed, it is the duty of the taxonomist to check whether the concepts are relevant to the taxonomy, they are used it, or it needs to be deleted from the concept list. The extracted results can be sorted using various parameters provided. There are different scores for help to decide whether a concept should be added or not from the corpus analysis to taxonomy. For instance, the MIS or the mutual information score can be such deciding factor when your are extracting the concepts, you can set a limit on the score you would allow the concepts to be taken from. The example dashboard looks like below (Fig. 6).

This is just the basic example of how the analysis would look like, if the relevance of the document is not close to the desired limit, it also shows a red bar to suggest that the analysis quality was not good. To build your own text corpus for text mining, there are factors to keep in mind, there are different tools which you can use but the topic on which you want your analysis to be done should be clear, and you should have you own thesaurus, from where it can search for the relevance of the document. The thesaurus should contain relevant details and concepts. And that is it so that you can enrich your taxonomy and ontology from text corpora.

7 Semantic Text Mining

In the previous section, we got the basic concepts of how text mining works, now we see it from the semantic and knowledge graphs perspective, semantic is the study of meaning of language, the meaning is the key term, it works on the relationship between word and phases, how the meaning or the context is present with it. Semantic text mining is just programmatic extraction of meaning from language, now as saw in the above sections, the text mining is just extracting the text and understanding the context, whereas in semantic text mining, it tries to copy human strategies for understanding. To deduce the meaning for the text, semantic text mining faces some of the same challenges as text mining as the major ones being disambiguation of the meaning. Homographs and nested terms with the help of synonyms do help in improving performance. Semantic works with understanding what the meaning of a sentence, paragraph or text means it can be focuses on the word and getting the exact meaning of that word, then the next word and how the context or the meaning is connected when it is used with the next word. The context of the sentence can be based on two key points, it can be depending on the local context where the context is about the word and which word is used or it can be the global context where the context can be used with the globally, it can be one word or sentence which is globally known or shared among everyone. For instance, sun rises in the east, it is universally accepted and known by everyone. There is also the third kind which is under the domain of local context, it is personal knowledge this is the meaning, or the context given by brain when a person reads a word. It can be different to different people bases on their preference and the surrounding environment. The global context is used to build the knowledge model, because it is something which is same for everyone and does not require to be changes based on perspective.

Knowledge model is used by machine to copy the human like strategies while analyzing a text and understanding its meaning. Knowledge model can contain different components like it can have domain-level-specific context. Where the information about the topic which it needs to build upon is stored, all the related concepts comes under the domain specific concept. Then comes the relationships between the concepts where related concepts are kept; additional context information is also an important component when it comes to building knowledge model. The additional context information helps with the alternate information where it can compare with the broader domain. Then comes the supplement knowledge model as the name suggests its purpose is to extend the knowledge with the usage example, text corpora come under this domain, it is major supplement for any knowledge model as it extends it domain and also gives various knowledge about the concepts. Other components can be co-occurrence of the concepts, relevance of the concepts of the domain, it also has information about the synonyms which increases the performance of the text corpus, and it can also include on a smaller level some of ontologies.

Semantic text mining uses context to determine the meaning of the individual terms or word or, the broader concept for the words to be in. The thing to be remember is semantic text mining works based on a global context, which is resolved using the knowledge model which has the information of the related concepts or the broader concepts.

8 Related Thing to Keep in Mind While Text Mining (State of the Art)

For querying the data, SPARQL query language is used; this is quite close to SQL but has some varied syntax which looks weird at first, but it falls on the hand after using it for some time. To give a brief overview, it already has the same select statement which selects the data according to the query. It also has key terms like ASK, construct and describe where ask just verifies the patter in which the query is called for, if the pattern matches with the query, then the statement returns "true" else it returns false. For instance: ASK { ?x name "Rohan"} it will search for data that has the name Rohan, it uses the triples terminology where subject, predicate, and object name and if the ontology has name "Rohan," it returns true, else it returns false. Describe form is used to return single result RDF graph containing the data about the resource. It can create a result set. For instance: DESCRIBE ?x WHERE {?x name "Rohan"} it will describe the Rohan set. CONSTRUCT returns a single RDF graph that is specified by the template, it results in the graph formed by each query solution in the result set of the query. Construct { something} where { ?x name "Rohan"}. This is quite helpful to visualize what data is there for the given topic. Some key points to keep in mind is that variables mostly starts with a "?" representing the query and then the statements are written. It also has some other keywords like order by, which is user for ordering the result, limit which is used for limiting the outcome of the result, filter which is used to filter the output based on the query for example it can be filter(?s = "Rohan"), i.e., it will filter out all the search results that has Rohan in it, or the keyword distinct which is used for the purpose to get the distinct values of the query output.

Data integration is an another crucial point for text mining as this is the basic need for the data mining, for the formats where we have the documents and the files related to the topic, it can be directly uploaded to PoolParty and corpus analysis as explained above can be done to get the concepts and the related concepts, it just needs some thesaurus which it can work on. It can have common data model RDF where the values are used from another of the RDF model; it can have common knowledge model where it is using the shared ontology or a public ontology already having the multiple inputs and is maintained by multiple sources; it depends on the knowledge model you are building for if that topic has shared ontology. You can have public ontologies and then you can align it to your use case whichever way it suits. Now, the thing to remember is that there are tools to convert data into the RDF triple format which can be found online. It is the mapping which is difficult, the mapping between the ontologies or when your are tailoring the ontology to suit the need and the data; once the ontologies mapping is done, you can easily wrap up the rest around it.

Linked data can also be used for enrichment of the taxonomy; these have huge potential and reduce the human effort to manually access the concepts. Linked data are data based on RDF so there is no need to convert it to RDF format; also, they are publicly available and there are plenty of sources online, though you should look for the licensed linked data resource or the trusted one which can be utilized. Linked data resources by default are stored on some ontologies tools to help build it. Also, there are machine-readable linked data at many sources like DBpedia; it is kind of Wikipedia but for ontologies. It has ontologies for geolocations which can be used if the data required is geographic, it all different sorts. There are sources like Worknet which are technically synonyms library. Always keep in mind for the linked data that it is best to connect it to source rather just keeping or storing it. So that any changes done on the source can be reflected back, unless you want it to be strictly static.

9 Resolving the Diverse Database Conflict

In the above sections, we read all about text mining and the basics for creating an ontology or a taxonomy, but it still does not solve the problem which we had from the start, i.e., data silos, how it can be removed the, answer is quite simple. We can build a central knowledge hub combining all the diverse database source into one. Actually, the more study it the idea get's simpler. If you are not utilizing the potential for sematic technologies, then the labor must be put into migrating all the data into a common platform where manually a database needs to create and then integrated into one giant compartment. Actually, with the help of data semantic, you can integrate all the different data sources without the need to actually migrate all these data sources into one and then query it. Though it still requires architecture components decisions and to check whether it is feasible for the organization or not. Even if it is not then from the moment it can be done so that the problem ends there and then and it is not dragged upon. The major advantage being that with increasing amount of data the data silos will not increase because of a centralized knowledge graph. With increasing variety, it is not necessarily needed to convert this into a certain form and it can be formatted and then stored in the database it can connect from. If the number of people producing and accessing the data is increasing, then the search access needs to be increased. Now the thing that needs attention is that you do not need to migrate your data into another source, your data by default is in the storage you are accessing from, and it is just integrated into the semantic metadata layer. Now you just need to query and search from the source which can be done using SPARQL [7] and Stardog. Stardog let us you query data with help of SPARQL, please find attached screenshot for the same (Fig. 7).

In PoolParty, it can be done using the semantic integrator where you can create a unified semantic metadata layer so that you can tie it together from different sources which are actually different database storage. This has separate content on the metadata layer which results in the change for the navigation logic. PoolParty has four-layer architecture which helps the semantic integrator. The four-layer architecture is explained below.

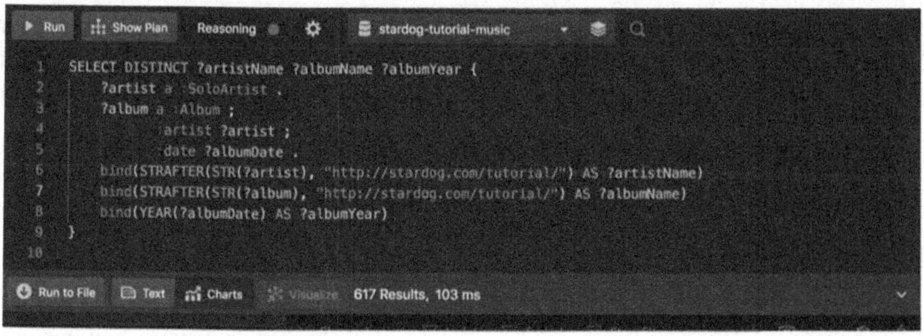

Fig. 7 SPARQL l query using Stardog

Navigation layer: This is for the user interface parameter, or from the usability perspective, it is about the visual parts where you can filter the information, aggregations and visualize the concepts used in the PoolParty. It also offers the graph search which can be used for searching the database; this is highly customizable layer. This is the solution for the query across different platform we discuss in the introduction section and also for the real-time visualization.

Semantic layer: This contains the information about the metadata, which concepts are there, what are the relationships between the keywords and other metadata fields; this is the layer which links the metadata and identifies what it does. This layer stiches all together so that other layer can work properly.

Metadata layer: This can be used for storing the information about the content and can have details like the reference documents or the author, which data it is uploaded. It works with the semantic layer and then exposes the data to the navigation layer. This can remove the problem of data silos as different database can be stitched together, even different ontologies, knowledge graphs, and triple stores can also be directly integrated as one. It can have additional search engine like the Solr or the Elastic search which can be utilized. The metadata layer has the performance even while working with the large amount of data because of added advantages like it can have new additional indices. For triple store, it can store and query knowledge graphs, tools like Stardog; GraphDB can be integrated to work for the advantages. It gives you the flexibility to have your choice of the data schema.

Content Layer: Content from different sources can be added into one like text document; this works as the base for the metadata layer as this is where the content is extracted from the metadata layer does not actually connect with the storage systems like RDS or the CMS. This is usually the layer which works under the hood to provide a unified view. This can work as an orchestration tools for the semantic integrator.

Semantic can have very vast usage. It can be used in the field of ETL or extract transform load tools, where it can schedule workflows and it consists of DPU's or individual data processing units (DPUs); each particular DPU has a defined task and they are executed at a scheduled interval of time one after other. The biggest advantage being it can be connected to the graph database and hence can be used for faster querying the data. Knowledge graphs also consists of nodes and edges and it can be pretty beneficial because instead of having to traverse the database once, it can work with the respected edge and fetch data using the nodes.

10 Conclusion

To conclude the learning for the above section, this chapter deals with how we can remove the major hurdles of industrial big data using the power of semantics; it also touch base on certain application of web semantics like text mining. It gives a brief idea about the taxonomy and the ontologies and how it can be used for different purposes. It explains the key concept of the RDF triples and the value for the linked data and how it can be used. With big data comes various challenges in the field, it tackles some of those challenges. It shreds lights upon topics such as knowledge graphs and the power of ontologies and taxonomies and how it can be used. It gives the overview of some of the tools like PoolParty and Stardog. It covers some of the advanced topics of the taxonomies and the usage and also outlines the difference between the ontologies and taxonomies. It also explains the use of text corpus and the corpus management system for knowledge graphs and how it can be used to enrich taxonomies and ontologies and solving the problem of querying the database using the graphs technique.

Further scope for this can be showing a uniform view with how we can integrate machine learning models while working real time with the knowledge graphs or using text corpus to increase the accuracy of the ML model. There is huge untapped potential when we talk about semantics and machine learning that is why tech giants are trying to use the advantage of semantics in the field of machine learning as the semantic and machine learning have the same core principal of making the machines learn. The chapter can also further extend to cover how to create a search engine and improve the search results based on the performance of the metadata layer and how it works or step-by-step guide for creating the new engine.

References

1. Saleh, I., El-Tazi, N.: Finding Semantic Relationships in Folksonomies. December 2018,https://doi.org/10.1109/WI.2018.00-92
2. Company, P., Vergara, M., Mondragon, S.: Contribution to Product Semantics Taxonomy. CINEI, Sec, E-12071
3. https://www.ontotext.com/knowledgehub/fundamentals/what-is-rdf-triplestore/
4. https://www.ontotext.com/knowledgehub/fundamentals/what-are-ontologies/#:~:text=An%20ontology%20is%20a%20formal,relationships%20that%20hold%20between%20them.&text=However%2C%20unlike%20taxonomies%20or%20relational,in%20a%20variety%20of%20ways.

5. https://www.dataversity.net/taxonomy-vs-ontology-machine-learning-breakthroughs/
6. Francesconi, E, Montemagni, S., Peters, W., Tiscornia, D.: Integration a Bottom-up and Top-Down Methodology for Building Semantic Resources for the Multilingual Legal Domain (2010) https://doi.org/10.1007/978-3-642-12837-06
7. https://www.w3.org/TR/sparql11-query/#tempatesWithBNodes

Machine Learning-Based Wireless Sensor Networks

Lipika Mohanty, Junali Jasmine Jena, Manjusha Pandey, Siddharth Swarup Rautaray, and Sushovan Jena

Abstract With the progression in machine learning techniques, many applications of it, in various fields, have emerged. One of such fields is wireless sensor networks (WSNs) where machine learning is applied for various purposes. It is very much important for a wireless sensor network to adapt and work in dynamic environment and to track the target correctly. Also it needs to detect attacks and faults early. Machine learning techniques provide various solutions to the issues that are faced by WSNs. This chapter discusses some recent applications of machine learning techniques, to different areas of WSNs as well as the challenges faced.

Keywords Machine learning · Wireless sensor networks · Intrusion detection · DoS attacks · Anomaly detection · Localization

1 Introduction

In the past few decades, human has developed so many advancements in the surrounding. This is only possible because human has the tendency to learn and to know more. Learning makes us improved version of ourselves. So to make our system better, we need to make it learn. Machine learning [1–4] is the process that

L. Mohanty · J. J. Jena (✉) · M. Pandey · S. S. Rautaray
School of Computer Engineering, KIIT (Deemed to be University), Bhubaneswar, Odisha, India
e-mail: junali.jenafcs@kiit.ac.in

L. Mohanty
e-mail: lipika.mohantyfcs@kiit.ac.in

M. Pandey
e-mail: manjushafcs@kiit.ac.in

S. S. Rautaray
e-mail: siddharthfcs@kiit.ac.in

S. Jena
Wipro Limited, WIPRO BHDC, Bhubaneswar, Odisha, India
e-mail: sushovanjena@gmail.com

M. Pandey and S. S. Rautaray (eds.), *Machine Learning: Theoretical Foundations and Practical Applications*, Studies in Big Data 87,
https://doi.org/10.1007/978-981-33-6518-6_6

helps to improve the machine in terms of making decisions. It makes the computing process more effective and trustworthy.

With the advancement of machine learning techniques, there are many areas such as natural language, image, medical data, advertising and human resources which uses machine learning techniques to solve their issues and to enhance the effects of the system. One of such fields is wireless sensor networks where machine learning solves a lot of problem and also it decreases the human interventions.

Wireless sensor networks (WSNs) [5–8] are one of the most rising technologies for some real-time applications due to its size, cost-effectiveness and ease of implementation. WSN composed of tiny nodes with sensing, estimation, wireless message passing capabilities. There could be various types of sensors that are fitted in the sensor nodes, such sensors are thermal sensors, acoustic sensors, chemical sensors, pressure sensors, weather sensors and optical sensors. Due to this variety, WSNs have incredible capabilities for creating significant applications which has its own individual uniqueness and necessities. Figure 1 shows the basic implementation of WSN.

This chapter discusses various applications of WSNs where machine learning techniques are used for solving the issues present in WSNs. Most of the works that have been discussed here are from few years back, i.e. from 2012 to 2020, so that readers could get a knowledge of the current decade works. Some of the challenges [9–11] that WSNs face are discussed in Sect. 1.1. In Sect. 2, some of the survey papers have been discussed. In Sect. 3, recent works of various applications of WSNs by using machine learning have been discussed. In Sect. 4, the conclusion of this chapter has been discussed.

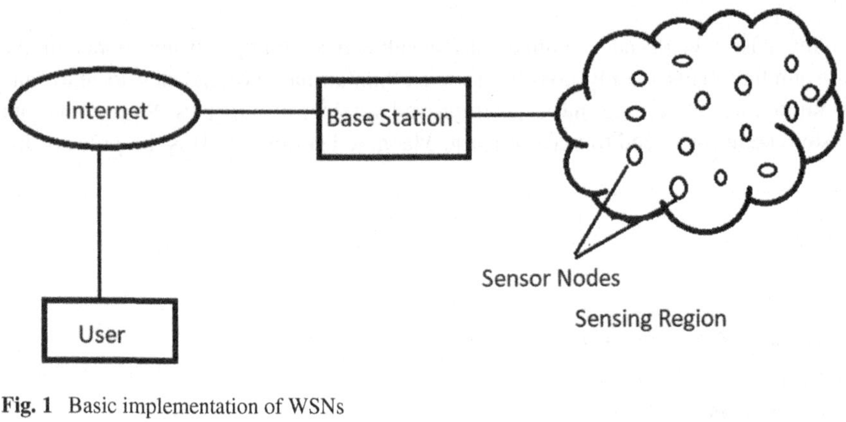

Fig. 1 Basic implementation of WSNs

1.1 Challenges

There are enormous issues associated with WSNs such as memory utilization of the nodes and the consumption of the energy; also, the geographical changes have an impact on WSNs. Here, we have listed few issues of WSNs. *Dynamic environment*: Wireless sensor networks need to monitor the environment as it is dynamic and changes rapidly over time. *Energy*: For different operations, WSNs require power consumption. Energy is consumed for performing various tasks such as collecting the data, processing the data and data communication. *Security*: It is very necessary for WSNs to protect the information which is communicated from one node to another or from sensors to the base stations; otherwise, information can be hacked. It is important for every sensor node and the base station that they should have the capability to identify that the received data is really sent by a genuine sender and not by any fraud that tricked the sensor nodes into accepting forged information. *Fault tolerance*: The network needs to work even if any failure occurs in any node. It should capable of adapting any change in connectivity. *Secure localization*: Locating a fault will require accurate location information in sensor networks. An intruder can easily change the unsecured location information. *Heterogeneity*: All the nodes in a group are not equal, they have different capability, i.e. some of the nodes have more potential than others. Cluster architecture is an example of where nodes form a cluster head and collect data from the node which is weak. *Synchronization*: To provide a common timescale for local clocks of sensor nodes in the wireless sensor networks, synchronization of time is required. It is very much important for the sensors to be synchronized with one another because asynchronous sensors may give inaccurate data estimation. Some of the synchronization protocols are having more precision but those protocols require more resources due to which there is loss of energy. So, based on the application, synchronization requires to be implemented appropriately.

2 State-of-the-Art Applications of Machine Learning in WSNs

Alsheikh et al. [12] presented an extensive literature survey of machine learning techniques which were used in WSN over a period of 2002–2013. They discussed the advantages and disadvantages of the machine learning methods in WSN and also gave a comparative study of the routing protocols distributed regression [13], sensor intelligence routing(SIR) [14], Q-map multicast [15], reinforcement learning-based geographic routing(RLGR) [16], Q-probabilistic [17] and feedback routing for optimizing multiple sinks (FROMS) [18]. In this paper, the some state-of-the-art machine learning techniques to solve the functional problems in WSNs like routing, localization of the dynamic target, clustering, aggregation of the data, processing of the queries and medium access control were reviewed. Gunduz et al. [19] reviewed how machine learning techniques can handle one of the major attacks on wireless sensor network, i.e. denial-of service (DoS) attack in each layer of TCP/IP protocol. Their main focus is DoS attacks at network layer and how machine learning methods solve them. Ahmad et al. [20] reviewed numerous papers which showed how machine learning strategies help to tackle most of the issues in wireless sensor networks. Mamdouh et al. [21] discussed many attacks on Internet of things (IoT) and WSN which can be handled by machine learning methods. Praveen et al. [22] discussed many machine learning algorithms with their pros and cons for WSN. Also they have discussed how machine learning techniques are used for synchronization, energy harvesting, mobile sink and congestion control. For intrusion detection and prevention, Chandre et al. [23] reviewed various machine learning and deep learning techniques used in WSNs. They had also done the performance analysis of the techniques. Dwivedi et al. [24] discussed some of the machine learning approaches for outlier detection in WSNs. Aghbari et al. [25] discussed some major challenges of WSNs; one of the challenges is to generate routing path while minimizing consumption of energy and maximizing life span of the network. They have surveyed papers where many optimization techniques are used to generate the routing paths from the starting node to the target node. Table 1 provides some recent review papers on WSNs using machine learning.

Table 1 Review papers on application of machine learning to WSNs

Sl. No.	Year	Application of machine learning	Name of the authors
1	2014	Wireless sensor networks [12]	Mohammad Abu Alsheikh, Shaowei Lin, Dusit Niyato and Hwee-Pink Tan
2	2015	Solutions to DoS attacks in wireless sensor networks [19]	Sedef Gunduz, Bilgehan Arslan and Mehmet Demirci
3	2017	Various areas of wireless sensor network [20]	Zaki Ahmad Khan, Abdus Samad
4	2018	Securing the internet of things and wireless sensor networks [21]	Marwa Mamdouh, Mohamed A. I. Elrukhsi and Ahmed Khattab
5	2019	Various areas of wireless sensor networks [22]	Praveen Kumar D., Tarachand Amgoth and Chandra Sekhara Rao Annavarapu
6	2019	Intrusion detection and prevention [23]	Pankaj R. Chandre, Parikshit N. Mahalle and Gitanjali R. Shinde
7	2020	Outlier detection in WSN [24]	Rajendra Kumar Dwivedi, Arun Kumar Rai, Rakesh Kumar
8	2019	Routing in WSNs [25]	Zaher Al Aghbari, Ahmed M. Khedr, Walid Osamy, Ifra Arif and Dharma P. Agrawal

3 ML in WSNs

To resolve the issues of WSNs, machine learning methods can be used as it can find the efficiency of the network on the given task by using the data and also it can predict the forthcoming efficiency.

3.1 Detection

From Table 1, it can be observed that machine learning can be used for various purposes in WSNs. One of the purposes of machine learning is to detect intrusion [26], spoofing and DoS attacks, to detect anomaly, to detect fault and leakage and also outlier detection in WSN which is discussed in this section. Table 2 provides the various works that have been done for various detections such as intrusion, fault, anomaly, leakage and outlier detection in WSNs using machine learning in the last few years.

Sandhya et al. [27] proposed a framework to detect intrusion with the help genetic *k*-means algorithm [28]. Chandre et al. [29] proposed a system which can detect intrusion using decision tree and also can prevent it. Geetha et al. [30] developed a system which can detect intrusion in WSNs and also identify the intrusive activities in the network by applying Naïve Bayes and Hoeffding rule-based decision tree. They also compared it with many other machine learning techniques and they found that it is giving much better faster result. Tan et al. [31] used random forest algorithm [32] to train the classifier and used synthetic minority oversampling technique (SMOTE) to balance the imbalance data set. de Lima Pinto et al. [33] built a system by using *k*-NN and *k*-means to detect the attacks. Shafiee et al. [34] used multi-layered neural network to detect the GPS spoofing attacks in GPS receivers. Abdulaziz et al. [35] proposed a system which uses decision tree [36] and support vector machine (SVM) [37] to detect the attacks.

Pachauri et al. [38] built a framework for anomaly detection by integrating random forest and additive regression (with *k*-NN) for classification and prediction, respectively. Jain et al. [39] built a machine learning framework which can detect anomalies using artificial neural network (ANN) [40, 41] and SVM [37]. Cauteruccio et al. [42] used unsupervised artificial neural network for identification of anomalies in heterogeneous wireless sensor networks and used multi-parameterized edit distance for could data analysis. Salem et al. [43] proposed a framework which uses SVM and linear regression model for anomaly detection in WSNs.

Warriach et al. [44] used hidden Markov model to detect faults in WSN. The data which was used to validate the proposed work was the real data collected from motes deployed in a living laboratory. Zidi et al. [45] used SVM [37] to create a decision function for the detection of fault in WSNs. Javaid et al. [46] developed four machine learning algorithms which are enhanced version of their traditional form. Those techniques are enhanced *K*-nearest neighbour (EKNN), enhanced extreme learning

Table 2 Application of machine learning in various types of detection in WSNs

Detection	Year	Purpose	Methodology used
Intrusion/spoofing/DoS attacks detection	2014	Intrusion detection [27]	Genetic *k*-means algorithm [28]
	2018	Intrusion detection [29]	Decision tree [36]
	2018	Intrusion detection [30]	Naïve Bayes [48] and Hoeffding rule-based decision tree [54]
	2019	Intrusion detection [31]	Random forest [32]
	2018	Spoofing attacks [33]	*k*-NN [55] and *k*-means [28]
	2019	DoS attack [35]	Decision tree and SVM
	2018	GPS spoofing attacks [34]	Multi-layered NN [56]
Anomaly detection	2015	Anomaly detection in medical WSN [38]	Random forest and additive regression (with *k*-NN)
	2014	Anomaly detection in medical WSN [43]	SVM and linear regression
	2016	Anomaly detection in smart city [39]	ANN [40, 41], SVM and multiclass SVM [57]
	2019	Anomaly detection in heterogeneous WSN [42]	ANN and multi-parameterized edit distance
Fault detection	2013	Fault detection in WSN [44]	HMM [58]
	2017	Fault detection [45]	SVM
	2019	Fault detection in WSNs [46]	EKNN, EELM, ESVM and ERELM
Leakage detection	2015	Leakage detection in fluid transportation system [47]	SVM, *k*-NN, Naïve Bayes and GMM [49]
	2019	Leakage detection in water pipeline [50]	SVM and PCA [59]
Outlier detection	2015	Outlier detection for WSN in harsh environments [51]	SVM
	2015	Outlier detection in WSN [52]	Least square SVM [53]

machine (EELM), enhanced support vector machine (ESVM) and enhanced recurrent extreme learning machine (ERELM) for detecting faults in WSNs. ERELM gave the best result out of those four techniques.

Sidra et al. [47] proposed a system which used SVM [37], *k*-NN, Naïve Bayes [48] and Gaussian mixture model (GMM) [49] for leak detection. Yang Liu et al. [50] used principal component analysis (PCA) for feature extraction and support vector machine (SVM) [37] for leakage detection.

Table 3 Machine learning techniques for target tracking in WSNs

Year	Purpose	Methodology used
2014	Target tracking in WSN [60]	Ridge regression [63] and vector-output regularized least squares
2018	Target tracking in WSN [64]	GRNN [65]
2018	Target localization and tracking in WSNs [66]	Regression tree
2020	Target tracking using signal strength in WSNs [67]	SVM and improved Kalman filter

Shahid et al. [51] used one-class SVM for outlier detection in wireless sensor networks in tough environment. Martins et al. [52] proposed a work in which they used least square SVM [53] for outlier detection in WSNs.

3.2 *Target Tracking*

Target tracking is one of the research areas in WSNs where the system needs to estimate the moving target's location instantly. In this section, some of the works done in this area using machine learning is discussed. Table 3 provides the various works done for tracking the targets in WSNs by using machine learning algorithms.

Mahfouz et al. [60] used machine learning with a Kalman filter [61, 62] for estimating the location of target instantly. They used kernel-based machine algorithms for the learning process. Those are ridge regression [63] and vector-output regularized least squares. Jhondale et al. [64] developed two novel algorithms, GRNN + KF and GRNN + Unscented KF, for target tracking in WSNs by using generalized regression neural network (GRNN) [65] and Kalman filtering framework. Ahmedi et al. [66] used regression tree algorithm for calculating the target's position. Wang et al. [67] proposed an innovative algorithm which combines the strength of SVM as well as an improved version of Kalman filter. The improved Kalman filter uses a threshold value which dynamically updates according to the target speed and network parameters so that the system remains stable.

3.3 *Localization*

As we know that determining the location of sensor nodes in wireless sensor networks is very tough work, machine learning can efficiently be applicable in localization in WSNs. In this section, some of the works that have already been done where machine learning methods are used for finding the physical location of the sensor nodes in

Table 4 Machine learning techniques for localization in WSNs

Year	Purpose	Methodology used
2013	Node localization in WSNs [68]	BPN [69] and RBF [70]
2018	Localization [71]	Regression model and SVM
2015	Localization in WSN [72]	ridge regression [63], SVR [73] and vector-output regularized least squares
2012	Target localization [74]	Ensemble SVR
2012	Range-free localization [75]	ANN

WSNs is discussed. Table 4 contains the various works done for localization in WSNs by using machine learning.

Singh et al. [68] proposed an algorithm which uses back propagation network (BPN) [69] and radial basis function (RBF) [70] network model which are artificial neural network models to solve the node localization problem. Bhatti [71] proposed a method which uses multiple regression model and SVM with RBF kernel to train as well as to make the training data set learn. Mahfouz et al. [72] used ridge regression [63], support vector regression (SVR) [73] and vector-output regularized least squares which are kernel-based machine learning algorithms. Similarly Kim et al. [74] used ensemble SVR for target localization, whereas Zheng et al. [75] used neural network for the same.

3.4 Security Enhancement

One of the foremost issues of wireless sensor networks is safety of the network which can cause the breakdown of the network. So in this section, some works that use machine learning to enhance the security of the WSNs are discussed. Aswathy et al. [76] used neural network and SVM for detecting DoS attacks in medium access control (MAC) layer. As DoS attacks are one of the major security flaws reason, early detection of DoS attacks can enhance the security. Yu et al. [77] also used machine learning technique for intrusion detection in WSNs.

3.5 Routing

For transmitting the information from source to destination, routing is very much important. It is one of the major challenges in WSNs to find the best routing policy in the presence of the problems such as unpredictable topology variations, energy constraints and link non-reliability. In this section, it has been discussed how machine learning helps in routing policy. Table 5 provides the various works done for routing protocol in WSNs by using machine learning algorithms.

Table 5 Machine learning algorithms for routing protocol

Year	Purpose	Methodology used
2007	Routing protocol in WSNs [17]	Reinforcement learning and Bayesian decision model
2016	Routing in WSNs [78]	SVM
2019	Robust routing [79]	CNN and LEACH-RP
2019	Routing algorithm for WSNs in IoT [80]	FBCFP

Rocio et al. [17] proposed a new routing algorithm, i.e. Q-probabilistic routing (Q-PR) [17], for WSNs by using two machine learning methods which are reinforcement learning and Bayesian decision model. Feeza et al. [78] used support vector machine algorithm to propose the protocol and that had been compared with LEACH protocol. Ali et al. [79] applied cyclic neural network (CNN) on LEACH routing protocol which effectively helped to reduce the data redundancy and resolve the issues. Thangaramya et al. [80] proposed a routing protocol based on neuro-fuzzy for WSNs, i.e. fuzzy-based cluster formation routing protocol (FBCFP). They compared this algorithm with LEACH, HEED [81] and FLCFP and they found that FBCFP gives better result than these three algorithms.

3.6 Clustering

Machine learning is also capable of forming the clusters in WSNs. Some of the works has been discussed in this section. Some of the works has been shown in Table 6 for clustering in WSNs using machine learning techniques.

Long Liu et al. [82] created a genetic algorithm [83] (GA)-based protocol implemented on LEACH [84], LEACH-GA, for forming the clusters in WSNs. The result showed that LEACH-GA protocol consumes optimal energy for WSNs. Zhang et al. [85] proposed a novel searching algorithm, peak searching algorithm, which is based on Bayesian optimization for clustering in WSNs. Jin Wang et al. [86] used a traditional learning algorithm, i.e. *K*-medoids and combined it with affinity propagation method for achieving good clustering performance. For clustering, Kotary et al. [87] used diffusion moth flame optimization [88]. The proposed work was compared

Table 6 Clustering in WSNs by machine learning

Year	Purpose	Methodology used
2011	Clustering in WSNs [82]	GA [83] and LEACH [84]
2017	Clustering in WSNs [85]	Peak Searching Algorithm
2019	Clustering in WSNs [86]	*K*-medoids [89]
2020	Distributed data clustering in WSNs [87]	Moth flame optimization [88]

with some nature-based optimization techniques such as diffusion particle swarm optimization (DPSO), distributed *K*-means (DK-Means), diffusion elephant herding optimization (DEHO) and diffusion whale optimization algorithm (DWOA).

4 Conclusion

In this represented chapter, we have noticed that machine learning techniques are capable to boost the capability of wireless sensor networks to become accustomed to the dynamic actions of its surrounding environment. Some of the issues of WSNs are also discussed in this chapter. Machine learning helps to solve those issues and to provide innovative solutions for detecting attacks, anomalies, faults, leakage. It is also gives better result in target tracking, localization, security enhancements and clustering in WSNs. However, some drawbacks are also there with machine learning algorithms as it is computationally costly and as WSNs has a framework with limited resources so it uses some amount of energy to estimate accurate prediction and to extract the general relationship with data set. Also gaining the knowledge through examples requires a lot of data set.

References

1. Michie, D., Spiegelhalter, D. J., & Taylor, C. C. (1994). Machine learning. *Neural and Statistical Classification, 13*(1994), 1–298.
2. Jordan, M. I., & Mitchell, T. M. (2015). Machine learning: Trends, perspectives, and prospects. *Science, 349*(6245), 255–260.
3. Alpaydin, E. (2020). *Introduction to machine learning*. MIT Press.
4. Dietterich, T. G. (2000). Ensemble methods in machine learning. In *International Workshop on Multiple Classifier Systems* (pp. 1–15). Berlin, Heidelberg: Springer.
5. Akyildiz, I. F., Weilian, Su., Sankarasubramaniam, Y., & Cayirci, E. (2002). Wireless sensor networks: A survey. *Computer Networks, 38*(4), 393–422.
6. Tree, S. (2014). Wireless sensor networks. *Self, 1*(R2), C0.
7. Karl, H., & Willig, A. (2007). *Protocols and architectures for wireless sensor networks*. Wiley.
8. Romer, K., & Mattern, F. (2004). The design space of wireless sensor networks. *IEEE Wireless Communications, 11*(6), 54–61.
9. Sharma, S., Bansal, R. K., & Bansal, S. (2013). Issues and challenges in wireless sensor networks. In *2013 International Conference on Machine Intelligence and Research Advancement* (pp. 58–62). IEEE.
10. Seah, W. K. G., Eu, Z. A., & Tan, H.-P. (2009). Wireless sensor networks powered by ambient energy harvesting (WSN-HEAP)-Survey and challenges. In *2009 1st International Conference on Wireless Communication, Vehicular Technology, Information Theory and Aerospace & Electronic Systems Technology* (pp. 1–5). IEEE.
11. Sharma, K., & Ghose, M. K. (2010). Wireless sensor networks: An overview on its security threats. In IJCA, Special Issue on "Mobile Ad-hoc Networks" MANETs (pp. 42–45).
12. Alsheikh, M. A., Lin, S., Niyato, D., & Tan, H.-P. (2014). Machine learning in wireless sensor networks: Algorithms, strategies, and applications. *IEEE Communications Surveys & Tutorials, 16*(4), 1996–2018.

13. Guestrin, C., Bodik, P., Thibaux, R., Paskin, M., & Madden, S. (2004). Distributed regression: an efficient framework for modeling sensor network data. In *Proceedings of the 3rd International Symposium on Information Processing in Sensor Networks* (pp. 1–10).
14. Barbancho, J., León, C., Javier Molina, F., & Barbancho. A. (2007). A new QoS routing algorithm based on self-organizing maps for wireless sensor networks. *Telecommunication Systems, 36*(1–3), 73–83.
15. Sun, R., Tatsumi, S., & Zhao, G. (2002). Q-map: A novel multicast routing method in wireless ad hoc networks with multiagent reinforcement learning. In *2002 IEEE Region 10 Conference on Computers, Communications, Control and Power Engineering. TENCOM'02. Proceedings* (Vol. 1, pp. 667–670). IEEE.
16. Dong, S., Agrawal, P., & Sivalingam, K. (2007). Reinforcement learning based geographic routing protocol for UWB wireless sensor network. In *IEEE GLOBECOM 2007-IEEE Global Telecommunications Conference* (pp. 652–656). IEEE.
17. Arroyo-Valles, R., Alaiz-Rodriguez, R., Guerrero-Curieses, A., & Cid-Sueiro, J. (2007). Q-probabilistic routing in wireless sensor networks. In *2007 3rd International Conference on Intelligent Sensors, Sensor Networks and Information* (pp. 1–6). IEEE.
18. Forster, A., & Murphy, A. L. (2007). FROMS: Feedback routing for optimizing multiple sinks in WSN with reinforcement learning. In *2007 3rd International Conference on Intelligent Sensors, Sensor Networks and Information* (pp. 371–376). IEEE.
19. Gunduz, S., Arslan, B., & Demirci. M. (2015). A review of machine learning solutions to denial-of-services attacks in wireless sensor networks. In *2015 IEEE 14th International Conference on Machine Learning and Applications (ICMLA)*. IEEE.
20. Khan, Z. A., & Samad, A. (2017). A study of machine learning in wireless sensor network. *International Journal of Computer Networks and Applications 4*(4), 105–112.
21. Mamdouh, M., AI Elrukhsi, M., & Khattab, A. (2018). Securing the internet of things and wireless sensor networks via machine learning: A survey. In *2018 International Conference on Computer and Applications (ICCA)*. IEEE
22. Kumar, D. P., Amgoth, T., & Annavarapu, C. S. R. (2019). Machine learning algorithms for wireless sensor networks: A survey. *Information Fusion 49*, 1–25.
23. Detection C., Pankaj R., Mahalle, P. N. & Shinde, G. R. (2020). Deep learning and machine learning techniques for intrusion detection and prevention in wireless sensor networks: Comparative study and performance analysis. *Design frameworks for wireless networks* (pp. 95–120). Singapore: Springer.
24. Dwivedi, R. K., Rai, A. K., & Kumar, R. (2020). Outlier detection in wireless sensor networks using machine learning techniques: A survey. In *2020 International Conference on Electrical and Electronics Engineering (ICE3)* (pp. 316–321). IEEE.
25. Al Aghbari, Z., Khedr, A. M., Osamy, W., Arif, I., & Agrawal, D. P. (2019). Routing in wireless sensor networks using optimization techniques: A survey. *Wireless Personal Communications* 1–28.
26. Alrajeh, N. A., Khan, S., & Shams, B. (2013). Intrusion detection systems in wireless sensor networks: A review. *International Journal of Distributed Sensor Networks, 9*(5), 167575.
27. Sandhya, G., & Julian, A. (2014). Intrusion detection in wireless sensor network using genetic *K*-means algorithm. In *2014 IEEE International Conference on Advanced Communications, Control and Computing Technologies* (pp. 1791–1794). IEEE.
28. Ghezelbash, R., Maghsoudi, A., & Carranza, E. J. M. (2020). Optimization of geochemical anomaly detection using a novel genetic *K*-means clustering (GKMC) algorithm. *Computers & Geosciences, 134*, 104335.
29. Chandre, P. R., Mahalle, P. N., & Shinde, G. R. (2018). Machine learning based novel approach for intrusion detection and prevention system: A tool based verification. In *2018 IEEE Global Conference on Wireless Computing and Networking (GCWCN)*. IEEE.
30. Geetha, S., Dulhare, U. N., & Sivatha Sindhu, S. S. (2018). Intrusion detection using NBHoeffding rule based decision tree for wireless sensor networks. In *2018 Second International Conference on Advances in Electronics, Computers and Communications (ICAECC)* (pp. 1–5). IEEE.

31. Tan, X., Shaojing, Su., Huang, Z., Guo, X., Zuo, Z., Sun, X., & Li, L. (2019). Wireless sensor networks intrusion detection based on SMOTE and the random forest algorithm. *Sensors, 19*(1), 203.
32. Liaw, A., & Wiener, M. (2002). Classification and regression by randomForest. *R News, 2*(3), 18–22.
33. de Lima Pinto, E. M., Lachowski, R., Eduardo Pellenz, M., Penna, M. C., & Souza, R. D. (2018). A machine learning approach for detecting spoofing attacks in wireless sensor networks. In *2018 IEEE 32nd International Conference on Advanced Information Networking and Applications (AINA)* (pp. 752–758). IEEE.
34. Shafiee, E., Mosavi, M. R., & Moazedi, M. (2018). Detection of spoofing attack using machine learning based on multi-layer neural network in single-frequency GPS receivers. *The Journal of Navigation, 71*(1), 169–188.
35. Al-issa, A. I., Al-Akhras, M., ALsahli, M. S., Alawairdhi. M. (2019). Using machine learning to detect DoS attacks in wireless sensor networks. In *2019 IEEE Jordan International Joint Conference on Electrical Engineering and Information Technology (JEEIT)* (pp. 107–112). IEEE.
36. Liu, W., Chawla, S., Cieslak, D. A., & Chawla, N. V. (2010). A robust decision tree algorithm for imbalanced data sets. In *Proceedings of the 2010 SIAM International Conference on Data Mining* (pp. 766–777). Society for Industrial and Applied Mathematics.
37. Cortes, C., & Vapnik, V. (1995). Support vector machine. *Machine Learning, 20*(3), 273–297.
38. Pachauri, G., & Sharma, S. (2015). Anomaly detection in medical wireless sensor networks using machine learning algorithms. *Procedia Computer Science, 70,* 325–333.
39. Jain, R., & Shah, H. (2016). An anomaly detection in smart cities modeled as wireless sensor network. In *2016 International Conference on Signal and Information Processing (IConSIP)* (pp. 1–5). IEEE.
40. Hassoun, M. H. (1995). *Fundamentals of artificial neural networks.* MIT Press.
41. Graupe, D. (2013). *Principles of artificial neural networks.* (Vol. 7). World Scientific.
42. Cauteruccio, F., Fortino, G., Guerrieri, A., Liotta, A., Mocanu, D. C., Perra, C., Terracina, G., & Vega, M. T. (2019). Short-long term anomaly detection in wireless sensor networks based on machine learning and multi-parameterized edit distance. *Information Fusion, 52,* 13–30.
43. Salem, O., Guerassimov, A., Mehaoua, A., Marcus, A., & Furht, B. (2014). Anomaly detection in medical wireless sensor networks using SVM and linear regression models. *International Journal of E-Health and Medical Communications (IJEHMC), 5*(1), 20–45.
44. Warriach, E. U., & Tei, K. (2013). Fault detection in wireless sensor networks: A machine learning approach. In: *2013 IEEE 16th International Conference on Computational Science and Engineering* (pp. 758–765). IEEE.
45. Zidi, S., Moulahi, T., & Alaya, B. (2017). Fault detection in wireless sensor networks through SVM classifier. *IEEE Sensors Journal, 18*(1), 340–347.
46. Javaid, A., Javaid, N., Wadud, Z., Saba, T., Sheta, O. E., Saleem, M. Q., & Alzahrani, M. E. (2019). Machine learning algorithms and fault detection for improved belief function based decision fusion in wireless sensor networks. *Sensors, 19*(6), 1334.
47. Rashid, S., Akram, U., & Khan, S. A. (2015). WML: Wireless sensor network based machine learning for leakage detection and size estimation. *Procedia Computer Science, 63,* 171–176.
48. Rish, I. (2001). An empirical study of the naive Bayes classifier. *IJCAI 2001 Workshop on Empirical Methods in Artificial Intelligence, 3*(22), 41–46.
49. Reynolds, D. A. (2009). Gaussian mixture models. *Encyclopedia of Biometrics, 741.*
50. Liu, Y., Ma, X., Li, Y., Tie, Y., Zhang, Y., & Gao, J. (2019). Water pipeline leakage detection based on machine learning and wireless sensor networks. *Sensors, 19*(23), 5086.
51. Shahid, N., Naqvi, I. H., & Qaisar, S. B. (2015). One-class support vector machines: Analysis of outlier detection for wireless sensor networks in harsh environments. *Artificial Intelligence Review, 43*(4), 515–563.
52. Martins, H., et al. (2015). A machine learning technique in a multi-agent framework for online outliers detection in wireless sensor networks. In *IECON 2015–41st Annual Conference of the IEEE Industrial Electronics Society.* IEEE.

53. Suykens, J. A. K., & Vandewalle, J. (1999). Least squares support vector machine classifiers. *Neural Processing Letters, 9*(3), 293–300.
54. Pfahringer, B., Holmes, G., & Kirkby, R. (2008). Handling numeric attributes in hoeffding trees. In *Pacific-Asia Conference on Knowledge Discovery and Data Mining* (pp. 296–307). Berlin, Heidelberg: Springer.
55. Tominaga, Y. (1999). Comparative study of class data analysis with PCA-LDA, SIMCA, PLS, ANNs, and *k*-NN. *Chemometrics and Intelligent Laboratory Systems, 49*(1), 105–115.
56. Piramuthu, S., Shaw, M. J., & Gentry, J. A. (1994). A classification approach using multi-layered neural networks. *Decision Support Systems, 11*(5), 509–525.
57. Mathur, A., & Foody, G. M. (2008). Multiclass and binary SVM classification: Implications for training and classification users. *IEEE Geoscience and Remote Sensing Letters, 5*(2), 241–245.
58. Jordan, M. I., Ghahramani, Z., & Saul, L. K. (1997). Hidden Markov decision trees. In *Advances in Neural Information Processing Systems* (pp. 501–507).
59. Karamizadeh, S., Abdullah, S. M., Manaf, A. A., Zamani, M., & Hooman, A. (2013). An overview of principal component analysis. *Journal of Signal and Information Processing, 4*(3B), 173.
60. Mahfouz, S., et al. (2014). Target tracking using machine learning and Kalman filter in wireless sensor networks. *IEEE Sensors Journal, 14*(10), 3715–3725.
61. Welch, G., & Bishop, G., (1995). An introduction to the Kalman filter.
62. Julier, S. J., & Uhlmann, J. K. (1997). New extension of the Kalman filter to nonlinear systems. In *Signal Processing, Sensor Fusion, and Target Recognition VI* (Vol. 3068, pp. 182–193). International Society for Optics and Photonics.
63. Zhang, Y., Duchi, J., & Wainwright, M. (2013). Divide and conquer kernel ridge regression. In *Conference on learning theory* (pp. 592–617).
64. Jondhale, S. R., & Deshpande, R. S. (2018). Kalman filtering framework-based real time target tracking in wireless sensor networks using generalized regression neural networks. *IEEE Sensors Journal, 19*(1), 224–233.
65. Specht, D. F. (2006). GRNN with double clustering. In *The 2006 IEEE International Joint Conference on Neural Network Proceedings* (pp. 5074–5079). IEEE.
66. Ahmadi, H., Viani, F., & Bouallegue, R. (2018). An accurate prediction method for moving target localization and tracking in wireless sensor networks. *Ad Hoc Networks, 70*, 14–22.
67. Wang, X., Liu, X., Wang, Z., Li, R., & Yiguang, Wu. (2020). SVM+ KF target tracking strategy using the signal strength in wireless sensor networks. *Sensors, 20*(14), 3832.
68. Singh, P., & Agrawal, S., (2013). TDOA based node localization in WSN using neural networks. In *2013 International Conference on Communication Systems and Network Technologies* (pp. 400–404). IEEE.
69. Jin, W., Li, Z. J., Wei, L. S., & Zhen, H. (2000). The improvements of BP neural network learning algorithm. In *WCC 2000-ICSP 2000. 2000 5th International Conference on Signal Processing Proceedings. 16th World Computer Congress 2000* (Vol. 3, pp. 1647–1649). IEEE.
70. Wilamowski, B. M., & Jaeger, R. C. (1996) Implementation of RBF type networks by MLP networks. In *Proceedings of International Conference on Neural Networks (ICNN'96)* (Vol. 3, pp. 1670–1675). IEEE.
71. Bhatti, G. (2018). Machine learning based localization in large-scale wireless sensor networks. *Sensors, 18*(12), 4179.
72. Mahfouz, S., Mourad-Chehade, F., Honeine, P., Farah, J., & Snoussi, H. (2015). Kernel-based machine learning using radio-fingerprints for localization in WSNs. *IEEE Transactions on Aerospace and Electronic Systems, 51*(2), 1324–1336.
73. Awad, M., & Khanna, R. (2015). Support vector regression. In *Efficient learning machines* (pp. 67–80). Berkeley, CA: Apress.
74. Kim, W., Park, J., Yoo, J., Kim, H. J., & Park, C. G. (2012). Target localization using ensemble support vector regression in wireless sensor networks. *IEEE Transactions on Cybernetics., 43*(4), 1189–1198.
75. Zheng, J., & Dehghani, A. (2012). Range-free localization in wireless sensor networks with neural network ensembles. *Journal of Sensor and Actuator Networks., 1*(3), 254–271.

76. Raj, A. B., Ramesh, M. V., Kulkarni, R. V., & Hemalatha, T. (2012). Security enhancement in wireless sensor networks using machine learning. In *2012 IEEE 14th International Conference on High Performance Computing and Communication & 2012 IEEE 9th International Conference on Embedded Software and Systems* (pp. 1264–1269). IEEE.
77. Yu, Z, Tsai J. J. (2008). A framework of machine learning based intrusion detection for wireless sensor networks. In *2008 IEEE International Conference on Sensor Networks, Ubiquitous, and Trustworthy Computing (SUTC 2008)* 2008 Jun 11 (pp. 272–279). IEEE.
78. Khan, F., Memon, S., & Jokhio, S. H. (2016). Support vector machine based energy aware routing in wireless sensor networks. In *2016 2nd International Conference on Robotics and Artificial Intelligence (ICRAI)* (pp. 1–4). IEEE.
79. Ali, B., Mahmood, T., Abbas, M., Hussain, M., Ullah, H., Sarker, A., & Khan, A. (2019). LEACH robust routing approach applying machine learning. *IJCSNS, 19*(6), 18–26.
80. Thangaramya, K., Kulothungan, K., Logambigai, R., Selvi, M., Ganapathy, S., & Kannan, A. (2019). Energy aware cluster and neuro-fuzzy based routing algorithm for wireless sensor networks in IoT. *Computer Networks, 151*, 211–223.
81. Younis, O., & Fahmy, S. (2004). HEED: A hybrid, energy-efficient, distributed clustering approach for ad hoc sensor networks. *IEEE Transactions on Mobile Computing, 3*(4), 366–379.
82. Liu, J.-L., & Ravishankar, C. V. (2011). LEACH-GA: Genetic algorithm-based energy-efficient adaptive clustering protocol for wireless sensor networks. *International Journal of Machine Learning and Computing, 1*(1), 79.
83. Maulik, U., & Bandyopadhyay, S. (2000). Genetic algorithm-based clustering technique. *Pattern Recognition, 33*(9), 1455–1465.
84. Fan, X., & Yulin, S. (2007). Improvement on LEACH protocol of wireless sensor network. In *2007 International Conference on Sensor Technologies and Applications (SENSORCOMM 2007)* (pp. 260–264). IEEE.
85. Zhang, T., Zhao, Q., Shin, K., & Nakamoto, Y. (2018). Bayesian-optimization-based peak searching algorithm for clustering in wireless sensor networks. *Journal of Sensor and Actuator Networks, 7*(1), 2.
86. Wang, J., Gao, Y., Wang, K., Sangaiah, A. K., & Lim, S.-J. (2019). An affinity propagation-based self-adaptive clustering method for wireless sensor networks. *Sensors, 19*(11), 2579.
87. Kotary, D. K., & Nanda, S. J. (2020). Distributed robust data clustering in wireless sensor networks using diffusion moth flame optimization. *Engineering Applications of Artificial Intelligence, 87*, 103342.
88. Mirjalili, S. (2015). Moth-flame optimization algorithm: A novel nature-inspired heuristic paradigm. *Knowledge-Based Systems, 89*, 228–249.
89. Sheng, W., & Liu, X. (2004). A hybrid algorithm for *k*-medoid clustering of large data sets. In *Proceedings of the 2004 Congress on Evolutionary Computation* (IEEE Cat. No. 04TH8753) (vol. 1, pp. 77–82). IEEE.

AI to Machine Learning: Lifeless Automation and Issues

Subhashree Darshana, Siddharth Swarup Rautaray, and Manjusha Pandey

Abstract Machine learning makes the computer able to perform without explicit programming. So, machine learning is now applied to each and every field of our daily life. The broad range of applications of machine learning are disease detection, weather forecasting, gaming, political discussion, business analytics, acoustics, agriculture, energy forecasting, genomics, etc. The advances in artificial intelligence and machine learning is a combination of tools and techniques used together to solve cognitive problems. The concepts have been effectively executed through BERT and the GPT-2 architectures. Convolutional neural network which implements depth-wise separable convolution and other neural networks are also used based on the requirement of the application. Machine learning strategies used for prediction and prognosis of the COVID-19 are partial derivative regression and nonlinear machine learning. Adaptive neural fuzzy inference system is used for wind power detection in power systems. Hence, it gives richer proposals and bits of knowledge for the ensuing decisions based on past information and activities with the extreme scope of production enhancement.

1 Introduction

Mitchell's popular machine learning book [1] starts with the explanation: "Ever since computers were designed, we have pondered whether they can be made to memorize. In case we manage to know how to program them to memorize or to move forward naturally with involvement - the outcome would be dramatic." This

S. Darshana (✉) · S. S. Rautaray · M. Pandey
School of Computer Engineering, KIIT (Deemed to be University), Bhubaneswar, Odisha, India
e-mail: subhashree.darshana1993@gmail.com

S. S. Rautaray
e-mail: siddharthfcs@kiit.ac.in

M. Pandey
e-mail: manjushafcs@kiit.ac.in

M. Pandey and S. S. Rautaray (eds.), *Machine Learning: Theoretical Foundations and Practical Applications*, Studies in Big Data 87,
https://doi.org/10.1007/978-981-33-6518-6_7

journey gave birth to a modern investigate region, i.e., machine learning, for computer science decades back. Till date, machine learning methods have been profoundly established in our each day's life, like reading news, handwriting detection, etc., when we are utilizing our cell-phones. Furthermore, machine learning has moreover picked up noteworthy accomplishments. For example, AlphaGO [2] vanquished human intervention within the diversion of GO, ResNet [3] outperformed human execution in image detection, Microsoft's speech recognition [4] approximated human level in discourse translation. Still these effective applications of machine learning are far away from completely robotized, i.e., "improving naturally from past experience."

Since there are no algorithms that can accomplish great execution on all conceivable learning problems with equal weight-age (concurring to no free lunch theorem [5, 6]), each angle of machine learning applications, such as highlight designing, model selection, and algorithm selection. Human experts are thus developing technology in machine learning applications. As these specialists are uncommon, the victory of machine learning comes at a incredible cost. Hence, automated machine learning (AutoML) does not fair stay an scholastic dream as mentioned in Michell's book, but moreover draws in more attention from research professionals. On the other hand if we will be able to take the human out of these machine learning applications, it ables us to empower quicker sending of machine learning algorithms over organizations, productively approve and benchmark the execution of conveyed arrangements, and make specialists center more on issues with more application and commerce values. These would make machine learning much more available for real-world utilizations, driving to unused levels of competence and customization, of which the affect can be without a doubt emotional. Propelled by the over scholarly dream and commonsense needs, in later a long time, AutoML has risen as a unused sub-area in machine learning. It has got more consideration not as it were in machine learning but moreover in computer vision, information mining, and normal dialect handling. Till date, AutoML has as of now been effectively connected in numerous vital problems.

2 Current State of Art

With all of the progressions made in artificial intelligence and machine learning nowadays, there appears to be an inundation of tools and techniques that are used to serve cognitive capabilities [8]. The current state of the craftsmanship in language preparing has viably blended repetitive and convolutional neural network,making modern strategies inside. All of this centers around transformers, self-attention and word embeddings. These concepts all offer assistance to show connections of words in greatly parallel CNNs or FCNs that can successfully get it full corpuses. A transformer performs according to its nomenclature and converts a set of word embeddings into another set of word embeddings or similar structures. It is especially successful for machine interpretation, word generation, or vector creation for classification. This is a profundity look at the transformers, but in brief they are fundamentally a compact of encoder and decoder systems that are prepared to achieve the assignments over. A capable concept that lies at the heart of the transformer is self-attention. Self-attention was as of late created by Google as a strategy to demonstrate the repetitive and spatial nature of dialects in a single arrange pass. To do this, complex scientific strategies are performed on each input grouping employing a inquiry, key, and esteem network. These values are converted to demonstrate spatial connections of all words in a sequential arrangement.

These concepts have been effectively executed through the BERT and the GPT-2 architectures. All the networks are differentiated in their own way; they have both executed an ultra-deep transformer demonstrate to achieve arrangement to arrangement and multi-task targets effectively. OpenAI, the maker of GPT-2, right now has models available for download that permit analysts and researchers to provoke the network and yield a imaginative "AI" composed text. There are various utilize cases for this in substance creation, chatbots, and content era. This accessibility of such a effective arrange ought to permit innovation to proceed its exponential development. Quote from a philosophically trained GPT-2 I created: "The meaning of life is to create the illusion of possibility for those who lack this ability…."

2.1 Computer Vision

It is an interdisciplinary scientific field which trains computers to interpret and understand the visual world. As convolutions have gotten to be more successful for image processing tasks, the community has settled on going more profound and more extensive in their organize models. This has brought the capable slant of residual networks. A residual neural network, such as within the case of the state of the craftsmanship Google Initiation organize, piece together residual blocks to construct the foremost capable CNNs. A leftover square is exceptionally commonly 2 or more parallel CNNs that are concatenated or included at the conclusion. This in turn leverages the effective parallel compute schemas to carry of superior outcomes than ever (Fig. 1).

At the same time, new strategies for convolutions have been created to require things encourage the research. A step over the Google Initiation Arrange is the Xception network. This network is about the same as its forerunner, but employments a modern sort of convolution called the depthwise separable convolution. The concept permits for much lower parameter counts and increased accuracy. It breaks the convolution operation into two parts. A depthwise convolution slides over each channel independently, and a point-wise convolution, pixel by pixel, slides over all channels. A culminating fad not just in Computer Vision, but over all utilize cases is exchange learning. The concept of transfer learning is notable due to the intensely prepared systems that have no have to be rehashed for each utilize case. Exchange learning regularly takes a pretrained feature extraction deficiting its classification layers and solidifies the weights. These are finally joined to the clients classifier as the input. The training algorithm trains the modern classifier to utilize the yields from the intensely prepared network.

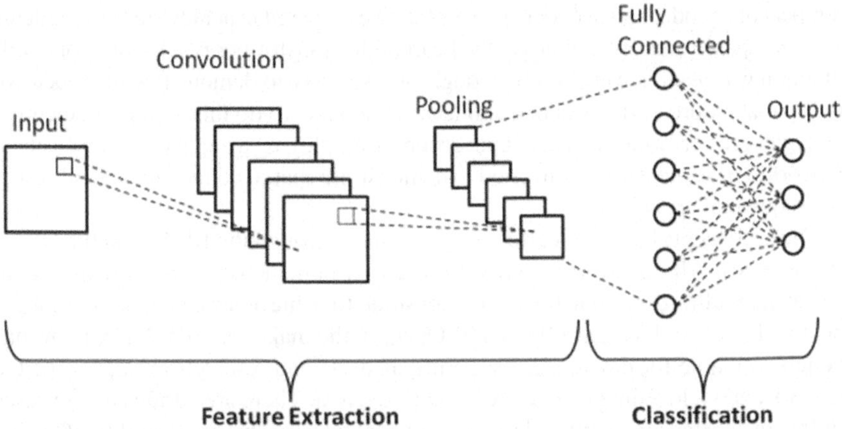

Fig. 1 CNN architecture

2.2 Generative Adversarial Networks

Generative adversarial networks are one of the foremost curiously shapes of AI nowadays. The concept is to construct a generator arrange that maps a vector into an picture network or comparative. This yield is encouraged to a discriminator organize which learns to translate the distinction between produced and genuine substance. As both systems are prepared in couple, the generator gets way better at fooling the discriminator because it gets way better as well. The ultimate result of this amusement hypothesis balance could be a generator able of making close genuine substance. This has lead to a few very astonishing craftsmanship, composing, and of course deep fakes.

2.3 Training and Deployment

AI has come to a point in which it is desirable and nearly basic to a few generation program frameworks. The strategies have permitted for worthy levels of accuracy to permit this innovation into the wild. So, it is exceptionally much appreciated to capable cloud computing technology that permits for adaptable arrangements and perfect preparing scenarios. In arrange to send effective models to the open, the ideal preparing and design is required. This is why Neural Architecture Search and Hyper Parameter tuning come into play. Utilizing intense levels of computing assets, Neural Architecture Search runs preparing on combinatorial level sets of potential models (layers, neurons, strategies, etc.). The outcomes can be utilized to send the ideal architectural model to train the signals and convey them to models. Here, a step encourage is taken to tune hyperparameters for an even more ideal model.

In 2020, library and cloud capabilities make it straightforward to send AI-fueled applications. The cloud can suit common shapes of AI and give it assets to scale and oversee application life cycles. Libraries, at the same time, have permitted for the information science and program advancement parts to partition. The Existence of generic model types made it required for the designers to get into the given library in about any programming platform to implement the data scientist's model.

3 From AI to ML

Artificial intelligence can be characterized as the capacity of a computer machine perform indicated capacities related with human being. It is fundamentally a reenacted adaptation of human insights where machines carry on like people and imitate their activities through programs. The fundamental concepts of artificial intelligence is capacity to rationalize and take phenomenal changes that have the most excellent chance of accomplishing a particular objective. Presently, a days counterfeit insights is utilized in wide range in ranges like space science health care, fund, mechanical technology, etc. The two major areas of AI are machine learning and deep learning. Machine learning can be named as an application of counterfeit insights which makes the framework able of autolearning and improves past experience of the dataset without interference of any explicit programming. It fundamentally centers on the adjustment of computer programs or logic switch able to access the information.

It makes a difference in automated analytical model building, serve to client by taking after neural network, statistics operation research and physics to discover the hidden up layer inside along with the data components without being unequivocally modified. A neural network can be defined as a bio-inspired computational model which works comparative to the working of a human anxious framework and comprises of interconnected units (neurons) and collect information from the outside inputs and hand-off the data between each unit. Essentially, it simulates lots of thickly interconnected brain cells which make us able to memorize things, recognize designs and make choices alike humans like Forecasting, algorithmic exchanging, securities classification, credit risk modeling and constructing proprietary indicators. A neural network contains layers of interconnected nodes. Each hub could be a perceptron and is comparative to a multiple linear regression.

Neural network consist of three layers: (a) input layer, (b) hidden layer and (c) output layer.

Each layer comprises of more than one hubs, spoken to within the graph by little circles. The lines between the hubs show the stream of data from one hub to another. The hubs of the input layer is detached; that is, they do not adjust the data. They get a single esteem on their input and copy the esteem to their different yields. The hubs of the covered up and output layer are dynamic, which implies they alter the information. The factors ($x1$, $x2$, … $x15$) hold the information to be evaluated. For example, they may be pixel values from picture, tests from an sound flag, etc. Each esteem from the input layer is copied and sent to all hidden nodes. This is known as completely interconnected structure. As appeared within the diagram, the values entering the covered up hub are duplicated by weight, a set of foreordained numbers put away in program. The weighted inputs are at that point included to create a single number. Before taking off the node, the number is passed through a nonlinear scientific work called sigmoid. There can be any number of layers and as many as number of nodes in each layer in a neural network. The size of the hidden layer is one-tenth the size of input layer.

4 Applications

Machine learning is one of the most interesting and demanding technology for researchers as well as common beings in the present days. As the nomenclature suggests, it aids the computer with intelligence and makes it similar to human beings. Nowadays, its been widely used for disease detection, weather forecasting, business analysis, stock market prediction, etc. Currently, with the exponential expansion of technology, the data size has increased and it is a challenge for researchers to deal with the data size and variety in data analysis algorithms also. So, the need of the hour has made it possible to induce machine learning into our day-to-day life.

4.1 COVID

The present worldwide flare-up of the novel corona virus infection 2019 (COVID-19) opened modern challenges for the investigate community. Machine learning (ML)-guided strategies can be valuable for highlight expectation, included hazard, and the causes of a practically equivalent to scourge. Such forecasts can be valuable for overseeing and interference the episode of such maladies. The first focal points of applying ML strategies are taking care of a wide assortment of information and simple identification of patterns and designs of an undetermined nature. In the present study, we propose a partial derivative regression and nonlinear machine learning (PDR-NML) strategy for worldwide widespread expectation of COVID-19 [19].

We utilized a progressive partial derivative linear regression model to hunt for the leading parameters within the dataset in a computationally productive way. Moreover, a nonlinear worldwide pandemic machine learning model was applied to the normalized highlights for making precise expectations. The outcomes conclude that the proposed ML strategy beated state-of-the-art strategies within the Indian populace and can too be a helpful instrument for making expectations for other countries. AI and ML innovation are utilized to improve the exactness of expectation for screening both irresistible and non-infectious illnesses [10]. The connection with well-being care starts with the advancement of the primary master framework called MYCIN created in 1976 [11]. MYCIN was planned to utilize 450 rules collected from a therapeutic expert to treat bacterial contamination by proposing anti-microbials to the patients. Such a master framework serves as clinical decision support for doctors and restorative specialists [12]. Present prospect of ML and AI innovation for the different widespread outbreak, it bolsters healthcare specialists in different communicable diseases (SARS, EBOLA, HIV, COVID_19) [13, 15, 17] and non-communicable infections (Cancer, Diabetic, Heart infection, and Stroke) [14, 16, 18] outbreak. The various ways how ML can be used for are as follows- (a) SARS-CoV-2 screening and treatment, (b) SARS-Cov-2 contact tracing, (c) SARS-CoV-2 prediction and forecasting and (d) SARS-CoV-2 drugs and vaccination.

Since the episode of the novel SARS-CoV-2, researchers and restorative businesses around the globe ubiquitously encouraged to battle against the widespread, looking elective strategy of fast screening and expectation prepare, contact following, estimating, and advancement of immunization or drugs with the more precise and solid operation. Machine learning and artificial intelligence are such promising strategies utilized by different healthcare suppliers. Here we addresses on recent consideration that apply such development innovation in increasing the analysts in numerous points, tending to the inconveniences and challenges, utilizing such calculation in helping restorative master in real-world issues [20].

This survey appears that the utilization of advanced innovation with AI and ML drastically makes strides the screening, forecast, contact following, determining, and drug/vaccine improvement with extraordinary unwavering quality. Larger part of the paper utilized profound learning calculations and is found to have more potential, strong and development among the other learning calculations. But, the current criticalness of the pandemic requires an made strides show with high-end execution exactness in screening and anticipating the SARS-CoV-2 with a distinctive kind of related malady by analyzing the clinical, mammographic, and statistic data of the suspects and contaminated patients. At last, it is evident that AI and ML can altogether make strides treatment, pharmaceutical, screening and forecast, determining, contact following, and drug/vaccine advancement for the COVID-19 widespread and decrease the human mediation in therapeutic hone. Be that as it may, most of the models are not sent sufficient to appear their real-world operation, but they are still up to the stamp to handle the widespread pandemic.

4.2 *Disease Detection*

Human health care is one of the most crucial aspect in the present time. Efficient disease or abnormality diagnosis can only make the healthcare system successful. When machine learning is induced into health care, it aims correct effective and robust disease detection and makes disease detection and gives ultimate care to patients. These disciplines confront the challenge of exploring modern methods, going past the conventional ones [7, 9].

Parkinson infection can be recognized productively utilizing ML. Background: administered machine learning has been proposed as a progressive approach for distinguishing delicate restorative picture biomarkers (or combination of them) permitting for programmed determination of person subjects. The point of this work was to survey the achievability of an administered machine learning calculation for the helped determination of patients with clinically analyzed Parkinson's illness (PD) and progressive supranuclear paralysis (PSP). Strategy: Morphological T1-weighted Attractive Reverberation Pictures (MRIs) of PD patients (20), PSP patients (20) and solid control subjects were utilized by a directed machine learning calculation based on the combination of Foremost Components Investigation as include extraction strategy and on Bolster Vector Machines as classifification calculation. The calculation was able to get voxel-based morphological biomarkers of PD and PSP [21].

Results: The algorithm permitted individual determination of PD versus controls, PSP versus controls and PSP versus PD with a precision, specificity and affectability >90%. Voxels influencing classification between PD and PSP patients included midbrain, pons, corpus callosum and thalamus, four basic locales known to be unequivocally included within the pathophysiological instruments of PSP. Comparison with existing strategies: Classification precision of person PSP patients was steady with past manual morphological measurements and with other directed machine learning application to MRI data, while precision within the discovery of person PD patients was significantly higher with our classification strategy. Conclusions: The calculation gives great separation of PD patients from PSP patients at a person level, in this way empowering the application of computer-based determination in clinical practices.

4.3 Wind Power Detection in Power Systems

Estimating of wind power, electric loads and vitality cost has gotten to be a major issue in control frameworks. Taking after needs of the showcase, different methods are utilized to figure the wind control, vitality cost and control demand. The issue confronted by control framework utilities is the inconstancy and non-schedulable nature of wind cultivate control era. These inborn characteristics of wind control have both specialized and commercial suggestions for efficient arranging and operation of control frameworks. To address the wind control issues, this paper presents the application of an adaptive neural fuzzy inference system (ANFIS) to exceptionally short-term wind determining utilizing a case ponder from Tasmania, Australia. The distinctive estimating issues and procedures within the range of electric stack estimating, vitality cost estimating and wind control forecast are said underneath.

A. Load forecasting

Load forecasting could be a process to anticipate stack for a future period. Application of stack determining falls into distinctive time skylines: long-term estimating (from one year to ten a long time), medium-term determining (from a few months to one year), short-term determining (from one-hour to one-week) and real-time or exceptionally short-term estimating (in minutes). Long-term figures influence the choices on era and transmission arranging, which is utilized for deciding the economical location, sort and measure of end of the control plants. Medium-term stack estimates are fundamental for era and transmission upkeep, conjointly for fuel planning. Precise short-term stack figures are essential for unit commitment a financial alacrity. Exceptionally short-term stack determining is for minutes ahead and is utilized for automatic generation control (AGC) [22, 23].

B. Price forecasting results

The cost estimating result displayed is based on past work on machine learning application to figure day-ahead utilizing NN model integrated with SD strategy [24]. Within the created cost forecast strategy, day-ahead power cost is gotten from the neural arrange that modifies the cost bends gotten by averaging chosen number of comparable cost days comparing to estimate day, i.e., two strategies were analyzed: (1) forecast based on averaging costs of comparable days and (2) expectation based on averaging costs of comparable days furthermore neural network refinement.

C. Wind power prediction result

A persistence model was too created for comparison. Determination is directly an industry benchmark for exceptionally short-term wind estimating and so is the foremost characteristic evaluation. The ANFIS show was created in a few diverse designs. The ANFIS show plan is adaptable and competent of taking care of quickly fluctuating information patterns [25].

4.4 *Agriculture*

Machine learning has developed with enormous information advances and high-performance computing to make unused opportunities for information seriously science within the multi-disciplinary agri-technologies space. In this paper, we show a comprehensive audit of inquire about devoted to applications of machine learning in agrarian generation frameworks. The works analyzed were categorized in: crop management, counting applications on abdicate forecast, infection discovery, weed discovery crop quality, and species recognition; livestock management, including applications on animal welfare and livestock production; water management; and soil management. The filtering and classification of the displayed articles illustrate how horticulture will advantage from machine learning innovations. By applying machine learning to sensor information, cultivate administration frameworks are advancing into genuine time artificial intelligence empowered programs that give wealthy proposals and bits of knowledge for agriculturist decision support and activity [26].

4.5 Politics

Issues dealing with varied opinion mining from microposts, and the challenges they force on an NLP framework, at the side an illustration application we have created to determine political leanings from a set of pre-election tweets. Whilst there are a number of estimation and analysis devices accessible which abridge positive, negative and impartial tweets approximately a given keyword or subject, these devices basically deliver poor outcomes, and work in a decently shortsighted way, utilizing as it were the nearness of certain positive and negative descriptive words as markers, or basic learning methods which don't work well on brief microposts [28]. On the other hand, intelligent devices which work well on motion picture and client audits cannot be utilized on microposts due to their brevity and need of setting. Our strategies make utilize of an assortment of modern NLP procedures in arrange to extricate more significant and higher quality suppositions and join extra-linguistic contextual data.

4.6 Genomics

The field of machine learning, which points to evolve computer algorithms that progress with involvement, holds guarantee to empower computers to help people within the analysis of huge, complex information sets. Here, we offer an outline of machine learning applications for the investigation of genome sequencing information sets, counting the annotation of grouping elements and epigenetic, proteomic or metabolite information [27]. We show contemplations and repetitive challenges within the application of supervised, semi-supervised and unsupervised machine learning methods, as well as of generative and discriminative displaying approaches [29]. We offer common rules to help within the determination of these machine learning strategies and their down to earth application for the investigation of hereditary and genomic information sets.

4.7 Networking

Machine learning (ML) has been relishing an exceptional surge in applications that unravel issues and empower robotization in differing spaces. Fundamentally, this is often due to the blast within the accessibility of information, noteworthy changes in

ML procedures, and headway in computing capabilities. Undoubtedly, ML has been connected to different unremarkable and complex issues emerging in arrange operation and administration. There are different studies on ML for particular zones in organizing or for particular arrange advances. This study is unique, since it together presents the application of assorted ML procedures in different key zones of organizing over distinctive arrange advances. In this way, perusers will advantage from a comprehensive talk on the diverse learning standards and ML methods connected to essential issues in organizing, counting traffic prediction, routing and classification, clog control, asset and blame administration, QoS and QoE administration and network security. Besides, this overview depicts the restrictions, donate experiences, investigate challenges and future openings to development ML in organizing. Hence, a convenient and periodic commitment of the implications of ML for networking, that's pushing the boundaries of autonomic network operation and management [31].

4.8 Energy Forecasting

Energy security and global warming are considered two of the most noteworthy challenges confronting the USA, whereas energy preservation is the elemental arrangement to both [31]. Owing to quick consumption of non-renewable assets, vitality productivity is one of the imperative issues that the world is confronting. Numerous nations are confronting vitality issues in all levels of their framework, industry and economy. The said worldwide vitality and natural challenges have driven city governments to continuously adjust their approaches, choices and techniques toward greener and vitality effective approaches. Within the USA, the state governments have set driven objectives in decreasing their greenhouse gas (GHG) outflows, such as 80% by 2050 in Unused York City and Boston. Other than understanding current utilization designs, determining urban buildings energy performance is vital to assembly such objectives.

5 Conclusion

Machine learning can be deployed and also can serve wide range of purposes, many of which have been mentioned. It gives richer proposals and bits of knowledge for the ensuing decisions based on past information and activities with the extreme scope of production enhancement. ML strategies have profoundly established in our lifestyle beginning from horticulture, malady discovery and forecast, legislative issues, control systems, energy framework and to each corner of our life. These days, machine learning methods are being broadly utilized to unravel real-world issues by storing, controlling, extricating and recovering information from expansive sources.

Administered machine learning methods have been broadly embraced in any case these methods demonstrate to be very expensive when the frameworks are actualized over wide extend of information. This can be due to the reality that significant sum of exertion and taken a toll is included since of getting huge labeled information sets. Thus, dynamic learning gives a way to reduce the labeling costs by labeling as it were the foremost useful instances for learning. So, in the forth coming days, it is anticipated that the utilization of ML models will be indeed more far reaching, permitting for the plausibility of coordinates and pertinent instruments. At the moment, all of the approaches respect individual approaches and arrangements and are not enough associated with the decision-making algorithms, as seen in other application spaces. This integration of computerized information recording, information examination, ML execution and decision-making or back will give down to earth tolls that come in line with the so-called knowledge-based agribusiness for expanding generation levels and bio-products quality. The performance of machine learning algorithms is more efficient for a classification assignment when the algorithms are combined. For the forecast of the correct output class, combined learner chooses the lesson to which most noteworthy likelihood has been assigned among all the learners.

References

1. Pereira, F., Mitchell, T., & Botvinick, M. (2009). Machine learning classifiers and fMRI: A tutorial overview. *Neuroimage, 45*(1), S199–S209.
2. Silver, D., Huang, A., Maddison, C. J., Guez, A., Sifre, L., Van Den Driessche, G., et al. (2016). Mastering the game of Go with deep neural networks and tree search. *Nature, 529*(7587), 484–489.
3. He, K., Zhang, X., Ren, S., & Sun, J. (2016). Deep residual learning for image recognition. In *Proceedings of the IEEE conference on computer vision and pattern recognition* (pp. 770–778).
4. Xiong, W., Droppo, J., Huang, X., Seide, F., Seltzer, M. L., Stolcke, A., et al. (2017). Toward human parity in conversational speech recognition. *IEEE/ACM Transactions on Audio, Speech, and Language Processing, 25*(12), 2410–2423.
5. Wolpert, D. H. (1996). The lack of a priori distinctions between learning algorithms. *Neural Computation, 8*(7), 1341–1390.
6. Wolpert, D. H., & Macready, W. G. (1997). No free lunch theorems for optimization. *IEEE Transactions on Evolutionary Computation, 1*(1), 67–82.
7. Krizhevsky, A., Sutskever, I., & Hinton, G. E. (2017). Imagenet classification with deep convolutional neural networks. *Communications of the ACM, 60*(6), 84–90.
8. https://towardsdatascience.com/the-state-of-ai-in-2020-1f95df336eb0.
9. Shortliffe, E. (Ed.). (2012) *Computer-based medical consultations: MYCIN* (Vol. 2). Elsevier.
10. Grottola, A., Marcacci, M., Tagliazucchi, S., Gennari, W., Di Gennaro, A., Orsini, M., et al. (2017). Usutu virus infections in humans: A retrospective analysis in the municipality of Modena, Italy. *Clinical Microbiology and Infection, 23*(1), 33–37.
11. Barbat, M. M., Wesche, C., Werhli, A. V., & Mata, M. M. (2019). An adaptive machine learning approach to improve automatic iceberg detection from SAR images. *ISPRS Journal of Photogrammetry and Remote Sensing, 156*, 247–259.
12. Mountrakis, G., Im, J., & Ogole, C. (2011). Support vector machines in remote sensing: A review. *ISPRS Journal of Photogrammetry and Remote Sensing, 66*(3), 247–259.

13. Shang, R., Qi, L., Jiao, L., Stolkin, R., & Li, Y. (2014). Change detection in SAR images by artificial immune multi-objective clustering. *Engineering Applications of Artificial Intelligence, 31,* 53–67.
14. Gao, F., You, J., Wang, J., Sun, J., Yang, E., & Zhou, H. (2017). A novel target detection method for SAR images based on shadow proposal and saliency analysis. *Neurocomputing, 267,* 220–231.
15. Colubri, A., Hartley, M. A., Siakor, M., Wolfman, V., Felix, A., Sesay, T., et al. (2019). Machine-learning prognostic models from the 2014–16 Ebola outbreak: Data-harmonization challenges, validation strategies, and mHealth applications. *EClinicalMedicine, 11,* 54–64.
16. Choi, S., Lee, J., Kang, M. G., Min, H., Chang, Y. S., & Yoon, S. (2017). Large-scale machine learning of media outlets for understanding public reactions to nation-wide viral infection outbreaks. *Methods, 129,* 50–59.
17. Nápoles, G., Grau, I., Bello, R., & Grau, R. (2014). Two-steps learning of Fuzzy Cognitive Maps for prediction and knowledge discovery on the HIV-1 drug resistance. *Expert Systems with Applications, 41*(3), 821–830.
18. Vickers, N. J. (2017). Animal communication: When I'm calling you, will you answer too? *Current Biology, 27*(14), R713–R715.
19. Lalmuanawma, S., Hussain, J., & Chhakchhuak, L. (2020). Applications of machine learning and artificial intelligence for Covid-19 (SARS-CoV-2) pandemic: A review. *Chaos, Solitons & Fractals*, 110059.
20. Kavadi, D. P., Patan, R., Ramachandran, M., & Gandomi, A. H. (2020). Partial derivative nonlinear global pandemic machine learning prediction of covid 19. *Chaos, Solitons & Fractals, 139,* 110056.
21. Salvatore, C., Cerasa, A., Castiglioni, I., Gallivanone, F., Augimeri, A., Lopez, M., et al. (2014). Machine learning on brain MRI data for differential diagnosis of Parkinson's disease and Progressive Supranuclear Palsy. *Journal of Neuroscience Methods, 222,* 230–237.
22. Senjyu, T., Mandal, P., Uezato, K., & Funabashi, T. (2005). Next day load curve forecasting using hybrid correction method. *IEEE Transactions on Power Systems, 20*(1), 102–109.
23. Hippert, H. S., Pedreira, C. E., & Souza, R. C. (2001). Neural networks for short-term load forecasting: A review and evaluation. *IEEE Transactions on Power Systems, 16*(1), 44–55.
24. Mandal, P., Senjyu, T., Urasaki, N., Funabashi, T., & Srivastava, A. K. (2007). A novel approach to forecast electricity price for PJM using neural network and similar days method. *IEEE Transactions on Power Systems, 22*(4), 2058–2065.
25. Osório, G. J., Matias, J. C. O., & Catalão, J. P. S. (2015). Short-term wind power forecasting using adaptive neuro-fuzzy inference system combined with evolutionary particle swarm optimization, wavelet transform and mutual information. *Renewable Energy, 75,* 301–307.
26. Liakos, K. G., Busato, P., Moshou, D., Pearson, S., & Bochtis, D. (2018). Machine learning in agriculture: A review. *Sensors, 18*(8), 2674.
27. Mackowiak, S. D., Zauber, H., Bielow, C., Thiel, D., Kutz, K., Calviello, L., et al. (2015). Extensive identification and analysis of conserved small ORFs in animals. *Genome Biology, 16*(1), 179.
28. Weichselbraun, A., Gindl, S., & Scharl, A. (2010). A context-dependent supervised learning approach to sentiment detection in large textual databases. *Journal of Information and Data Management, 1*(3), 329–329.
29. Libbrecht, M. W., & Noble, W. S. (2015). Machine learning applications in genetics and genomics. *Nature Reviews Genetics, 16*(6), 321–332.
30. Boutaba, R., Salahuddin, M. A., Limam, N., Ayoubi, S., Shahriar, N., Estrada-Solano, F., & Caicedo, O. M. (2018). A comprehensive survey on machine learning for networking: Evolution, applications and research opportunities. *Journal of Internet Services and Applications, 9*(1), 16.
31. Fathi, S., Srinivasan, R., Fenner, A., & Fathi, S. (2020). Machine learning applications in urban building energy performance forecasting: A systematic review. *Renewable and Sustainable Energy Reviews, 133,* 110287.

Analysis of FDIs in Different Sectors of the Indian Economy

Parikshit Barua, Sandeepan Mahapatra, Siddharth Swarup Rautaray, and Manjusha Pandey

Abstract Foreign direct investment (FDI) is the investment made by a foreign person/company to establish a business in another country. It is an essential factor for evaluating the progress of any country and plays a huge role in its economy. An increase in FDI means an increase in technology, infrastructure, employment, which in turn results in higher tax generation for the host country. It also leads to an increase in foreign reserves. With India's Prime Minister, Narendra Modi's $5 trillion economy target for India, his government has initiated various campaigns such as "Make in India," which would encourage joint ventures by foreign entities into India and further better the economy. This paper explores the trends of FDIs in various sectors of the Indian economy for the past 19 years and determines which sectors have seen a decrease in investments over the years, and hence, may need more funds from the government or ease of regulations to promote FDIs. Also, finding sectors that are being invested in the most will help us determine the strong sectors of the Indian economy. Using machine learning and forecasting, this project proposes a model to estimate the expected trends in FDIs which can possibly be seen in the coming few years. This will help the government make a decisive plan and form a budget for the upcoming financial years to cut the slowdown that the country is currently facing and make the economy grow further.

Keywords AR · Autoregression · ARIMA · Economy · FDI · MA · Moving average · SARIMA

P. Barua (✉) · S. Mahapatra · S. S. Rautaray · M. Pandey
School of Computer Engineering, KIIT (Deemed to be University), Bhubaneswar, Odisha, India
e-mail: parikshitb04@gmail.com

S. Mahapatra
e-mail: mahapatrasandeepan@gmail.com

S. S. Rautaray
e-mail: siddharthfcs@kiit.ac.in

M. Pandey
e-mail: manjushafcs@kiit.ac.in

M. Pandey and S. S. Rautaray (eds.), *Machine Learning: Theoretical Foundations and Practical Applications*, Studies in Big Data 87,
https://doi.org/10.1007/978-981-33-6518-6_8

1 Introduction

A foreign direct investment (FDI) is an investment made by a firm or individual in one country into business interests located in some other country. A person/firm establishing foreign business operations or acquiring foreign business assets in a foreign company is an instance of an FDI. Foreign direct investments (FDI) are a major monetary source for economic development of India as it reduces unemployment, creates job opportunities, increases the use of technology and managerial skills, removes balance of payment constraints, promotes exports, and generates a competitive environment and many other progressive outputs that differentiates this kind of investment from other funding sources.

Due to the lack of enough domestic capital for economic development, India needs foreign capital. India has become an investment hub over the last decade. The major areas of FDIs are—oil, mining, coal and gas, banking, insurance, transportation, finance, manufacturing, retailing, etc. FDI has been playing a major role in the growth of the Indian economy. Historically, India has followed a very cautious and selective approach regarding foreign capital, but after economic reforms in 1991, it has liberalized foreign direct investment policies. A number of measures were undertaken to promote FDI, and thus, the Government of India (GOI) has been successful in attracting more FDI in India. From the years 1991–1992 to 2011–12, India has fetched 4,26,318 US $ million FDI inflows (considering estimates made by RBI for the year 2010 to 2012).

This research paper explores the trends of FDIs in various sectors of India's economy in the past 19 years. The data (source: *dipp.gov.in*) consists of the records of FDIs in various sectors like agricultural machinery, electrical equipment, electronics, hospitals and diagnostic centers, etc., for different years starting from 2000 up to 2019. Graphical analysis based on the top 5 sectors has been done which shows that the service sector has reached its height in recent years. Most sectors are also showing an increasing trend except for a few sectors. It also goes through the impact of demonetization by the present ruling government in India from where we come to know that there has been a significant growth in the fields of computer software and hardware, telecommunications, and trading. The overall interpretations have been done using exploratory data analysis and forecasting techniques in the python framework. This report also helps to monitor and analyze the potential of inflow to India so that adequate measures can be taken for the improvement of the economy.

This paper is presented in five sections where Sect. 1 was an introduction to the paper and also a discussion about the technologies used. The state of the art has been presented under Sect. 2 of this paper. Section 3 discusses the model used and gives a detailed summary about it. Section 4 contains observations from the data and also the forecasting done by the model. The paper has been wrapped up and concluded in Sect. 5.

1.1 Technologies Used

This research work analyzes foreign direct investment (FDI) with the help of Python programming language. Python is a fairly simple programming language to learn, and it is easily readable in the sense one can tell what a line of code is doing just by looking at it, without any prior knowledge of its syntax. It contains various packages and modules with functions in them which makes the task of a data scientist very easy.

Firstly, the data set has been imported using Pandas, which is a library for manipulating data frames in the Python programming language. Pandas is very efficient to read, manipulate, and operate on tabular data. It treats each column of the table as an individual "Pandas Series," and it is very easy to perform calculations or apply functions to rows or columns of a table. Pandas has also been used for data pre-processing and has been used for transforming raw data into an understandable format.

Auto regressive integrated moving-average (ARIMA) model has been used for forecasting. Forecasting is the process of predicting the future based on past and present trends, and it is commonly used for predictions on time-series data. The ARIMA is actually a class of models that "explain" a given time series based on its own past values. It is a hybrid model made up by integration of two independent models, namely the auto regressive (AR) model and moving-average (MA) model. The stats models library provides the capability to fit an ARIMA model.

Finally, Python libraries like Matplotlib and Seaborn, which are excellent libraries for graphing and plotting in Python, have been used to perform data visualization which helps us look into the trends in various sectors of the economy and how the government has performed. Library like Numpy has also been used for faster computations.

2 State of the Art

This section of the paper work discusses the work done by other authors in this field. The author's name, title of his/her work, the proposal of the paper, and its findings or reviews have been given in a tabular format in Table 1.

3 ARIMA Model for Forcasting

The ARIMA model can be thought of to be a hybrid model made up of the integration of two simple models. The first simple model being the AR model.

Table 1 Tabular view of current research work carried out in the field

Author's name	Title	Proposal	Reviews
Noheed Khan 2017 [1, 9–16]	Impact of FDI and export on economic growth: Evidence from Pakistan and India	To find the relationship between FDI and economic growth	FDI and export have a positive relationship with economic growth
Wang Shaobin 2011 [2, 17]	Analysis on relationship between FDI and economic growth in ShaanXi based on OLS model	To find the relationship between FDI and economic growth	FDI and GDP have a positive correlation Although foreign direct investment is not all factors to the economic growth of Shaanxi, its role is still very large
Jinsheng Zhu 2011 [3]	The impact of FDI on China's economy security: An empirical analysis based on factor analysis and GRNN neural network	To find the impact of FDI on China's economic security	FDI sources, FDI in regional structure, and FDI in the industrial structure have a great effect on China's economic security
Yan Chen 2014 [4, 18–20]	The analysis of the technical progress effect of FDI in China	To find the relation between FDI and technological development	FDI has a promoting effect on China's technological progress
Jun Guan 2010 [5]	The trade effect analysis of foreign direct investment in Hebei	To find the effect of FDI in trades	It has a direct relationship with trade, thus having an increasing trend
Cheng Li-wei 2007 [6, 21–23]	Empirical investigation of FDI impact on China's employment	To find the impact of FDI over employability	It has been observed that FDI promotes employability
Yu Chao 2011 [7, 24, 25]	Effects of China's outward FDI on domestic employment	To examine the effect of outward FDI on domestic employment	China's outward FDI has a positive effect on domestic employment, the effect is not significant
V. Aruna 2012 [8]	Foreign direct investment (FDI) in the multi-brand retail sector in India—A boon or bane	To examine the negative and positive impacts on the entry of FDI in India	Experience of the last decade shows that small retailers have flourished in harmony with large outlets

3.1 Model 1: AR MODEL

The auto regressive (AR) model regresses the values of the time series against previous values of the same time series.

The equation of a simple AR model, AR ($p = 1$) is given by (1):

$$y_t = a_1 \cdot y_{t-1} + e_t \tag{1}$$

The value of the series at time "t" is directly proportional to its value at the previous step. e_t is the shock term or the white noise. Each shock is random and is not related to other shocks in the series.

The second simple model is the MA model.

3.2 Model 2: MA MODEL

The moving average (MA) model regresses the values of the time series against previous shock values of the same time series.

The equation of a simple MA model, MA ($q = 1$) is given by (2):

$$y_t = m_1 \cdot e_{t-1} + e_t \tag{2}$$

Combining both the AR and MA models gives us the ARMA model.

3.3 Hybrid Model 1: ARMA MODEL

The ARMA(AR + MA) is a combination of both the AR and MA models. This model regresses the time series against both the previous values and the previous shock terms.

A simple ARMA ($p = 1, q = 1$) model is given by Eq. (3):

$$y_t = a_1 \cdot y_{t-1} + m_1 \cdot e_{t-1} + e_t \tag{3}$$

The ARMA model can be extended to the ARIMA model by providing it with a parameter "d," which defines the number of "differencing" required by the series before fitting it to the model.

3.4 Hybrid Model 2: ARIMA MODEL

The autoregressive integrated moving average (ARIMA) is one of the most widely used models for forecasting. It uses both autoregressive (AR) and moving-average (MA) elements.

It requires "Stationary" time-series data for forecasting. A "Stationary" time-series data is one which follows the following conditions:

Condition 1: Has constant mean or the trend in the data is 0.

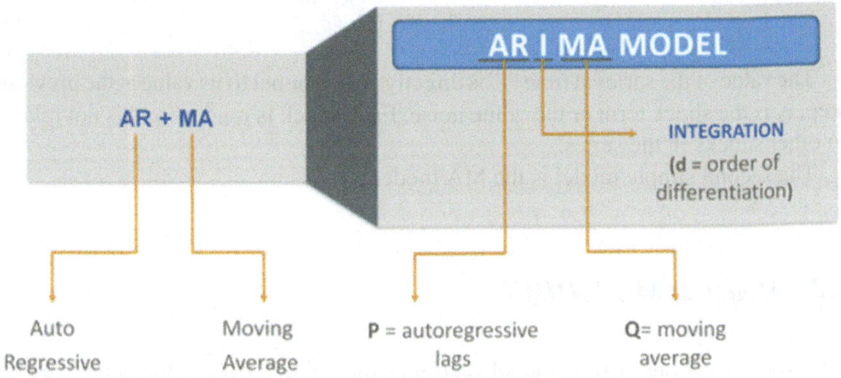

Fig. 1 ARIMA Model

Condition 2: Has constant variance.
Condition 3: Has autocovariance that does not depend on time.

There are various ways to transform a non-stationary data into stationary. One way is to take difference of the series.

If y denotes the *d*th difference of the series *Y*, then:

- If $d = 0, y_t = Y_t$
- If $d = 1, y_t = Y_t - Y_{t-1}$
- If $d = 2, y_t = (Y_t - Y_{t-1}) - (Y_{t-1} - Y_{t-2})$ …. and so on

An ARIMA model can be classified as "ARIMA(*p*, *d*, *q*)," where:

p: number of autoregressive terms or autoregressive lags
d: number of differences needed for stationarity
q: the number of moving-average terms

Figure 1 describes an ARIMA Model. Although the ARIMA model is capable of catching the trends in univariate data, it does not particularly succeed in catching the seasonality aspect of time-series data. ARIMA expects the data to be non-seasonal or does better with data that has the seasonal component removed.

4 Implementation and Results

This paper is aimed at analyzing the trends of FDIs in India over the last 20 years. It also tries to propose a model for forecasting future trends for FDIs in the country.

The data set has been prepared by collecting the data provided by the Government of India's site https://data.gov.in. (*Source:* https://data.gov.in/catalog/foreign-direct-investment-fdi-equity-inflows?filters%5Bfield_catalog_reference%5D=839421&format=json&offset=0&limit=6&sort%5Bcreated%5D=desc).

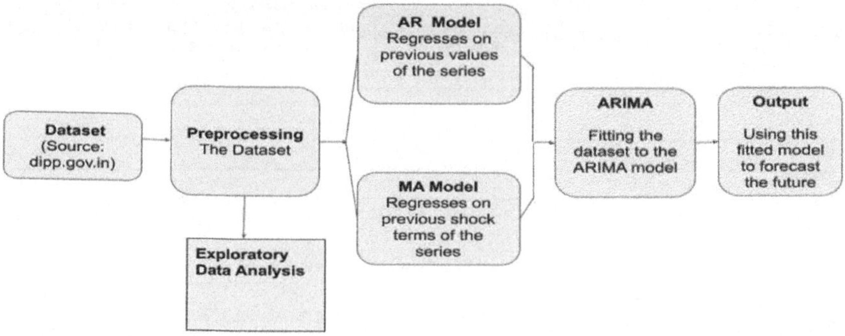

Fig. 2 Analytical life cycle for FDI

Figure 2 explains the work-flow and the approach that has been taken. Two data sets have been imported for this work. One containing the sector-wise FDI rates starting from the year 2000–01 up to 2016–17, and the second data set contains the total inflow in each sector starting from 2000 to 2001 to the last financial quarter of this financial year (2019–20). These two data sets have been joined on the sectors. A new column has been made which contains the difference of total inflow in each sector (till last quarter) and the sum of inflow till 2016–17. This gives the inflow from 2017 to last quarter. It has been assumed that two-third of this difference was the inflow for the year 2017–18, and the rest one-third has been the inflow for the year 2018–19 (GDP growth in 2017–18 was 7.2% compared to 6.8% in 2018–19).

This has been done in order to get an organized data set containing the inflow of FDI for each sector in the economy, starting from financial year 2000–01 up to financial year 2018–19. The data has been manipulated in such a way so that it is easy to work with and plot it. This data has been used to plot various graphs and draw meaningful conclusions as we shall see later on. This is also the type of data that a forecasting model expects.

Figure 3 shows the last five rows and all the features of the data set thus obtained.

Figure 4 is a visual representation of the FDIs in different sectors/industries of India for the past two decades. We can see the trend of FDI in all sectors of the economy. Some sectors have had huge fluctuations in FDI rates every year, whereas some sectors have received quite stable inflow of FDI over the years. FDI rates in few sectors have grown exponentially; while in some sectors, it has been very low since the beginning of this century.

Not much can be inferred from this plot. So, looking at the top five sectors which have been bringing in the most capital would be more meaningful. A plot of the top five sectors with the highest average investment over the past 19 years shows us the strong sectors or backbone of India's economy in terms of the rate of FDI.

Figure 5 shows the top five sectors (services, computer software, construction development, telecommunications, and automobile) that the government should primarily focus on maintaining and progressing as these are bringing in a lot of joint ventures and capital into the economy. Services sector has grown a lot whereas the

Sector	METALLURGICAL INDUSTRIES	MINING	POWER	NON-CONVENTIONAL ENERGY	COAL PRODUCTION	PETROLEUM & NATURAL GAS	BOILERS AND STEAM GENERATING PLANTS	PRIME MOVER (OTHER THAN ELECTRICAL GENERATORS)	ELECTRICAL EQUIPMENTS	COMPUTER SOFTWARE & HARDWARE	...	TRADI
2015-01-01	359.340000	684.390000	707.04	615.950000	0.000000	1079.020000	1.33	230.700000	574.830000	2296.04	...	2727.960(
2016-01-01	456.310000	520.670000	868.80	776.510000	0.000000	103.020000	77.91	159.130000	444.880000	5904.36	...	3845.320(
2017-01-01	1440.180000	55.750000	1112.98	783.570000	0.000000	180.400000	53.91	286.880000	2230.690000	3651.71	...	2338.400(
2018-01-01	694.446667	252.926667	1944.28	1921.053333	-0.006667	110.653333	45.48	276.453333	1075.433333	9872.94	...	6632.433:
2019-01-01	347.223333	126.463333	972.14	960.526667	-0.003333	55.326667	22.74	138.226667	537.716667	4936.47	...	3316.216(

5 rows × 63 columns

Fig. 3 Tail of the data set

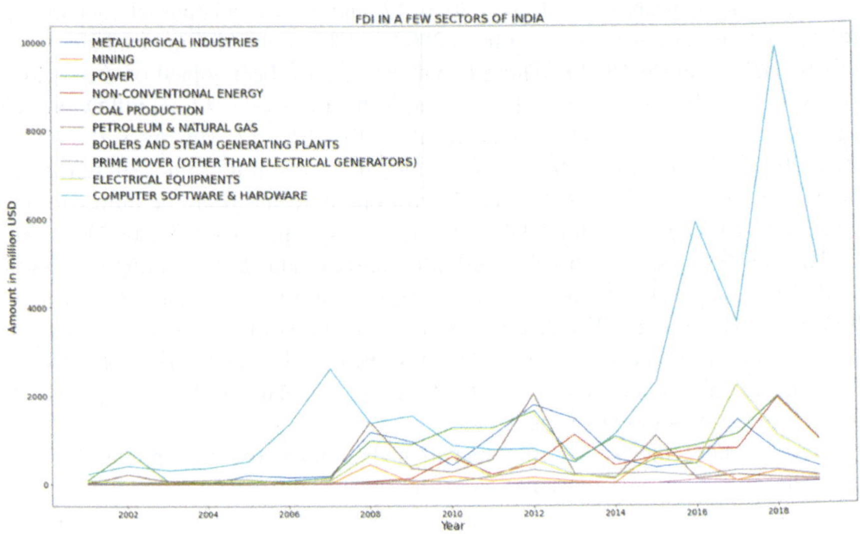

Fig. 4 FDIs in all sectors of Indian economy

construction development sector had its peak around 2009 and then started declining rapidly. Computer software and hardware, telecommunications, and construction development sectors have seen an uptrend since the start of this century.

The impact of FDIs into a particular sector depends on the government. Different sectors flourish over different governments. A histogram of the total amount of the inflow of capital before 2014 and after 2014 (Narendra Modi takes over as the Prime Minister of India) (Fig. 6) gives a clear picture of the changes in FDIs in different sectors. Services sector has remained unaffected and has been the sector receiving

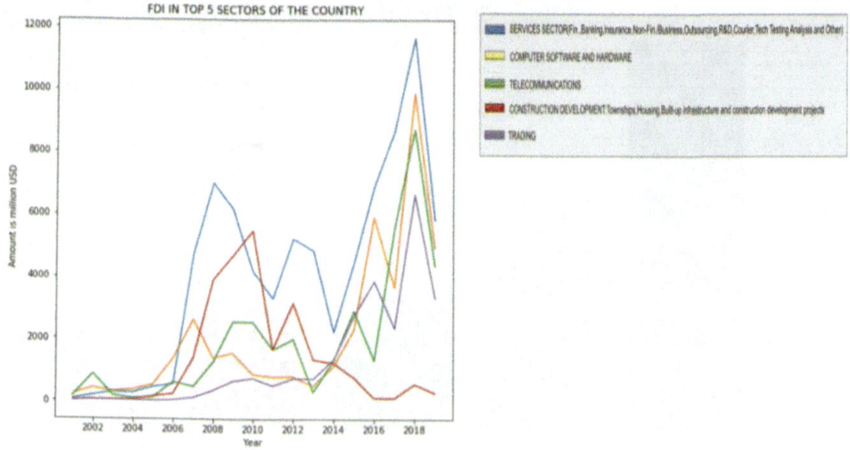

Fig. 5 Top 5 sectors in terms of FDI

most inflow both before and after 2014. Construction development sector has seen a decline under the current government and computer service, and hardware sector has flourished. Telecommunications sector too has been steadily growing under different governments. Drugs and pharmaceuticals sector has seen a decline since 2014 but trading sector has grown. Automobile sector has flourished since 2014.

4.1 Box–Jenkins Method

Using the Box–Jenkins method (Fig. 7) to find the best fit for a model that would be able to forecast future trends of FDI in any of the sectors, while focusing only on one sector ("Services" sector) to fit and tune the model.

The Box–Jenkins method is a common and systematic way of fitting and checking the model performances of time series models such as ARIMA. It is an iterative process consisting of four phases, namely "*identification*" of the time-series model and data, "*estimation,*" "*model diagnostics,*" and "*production,*" provided the "model diagnostic" curves are satisfactory. Following these steps makes sure that the best model and model parameters are chosen before the deployment of the model.

4.1.1 Identification

This is the first step of a series of steps that has been followed to build a model. A plot of the FDI into the services sector with respect to years (Fig. 8) gives an idea about the stationarity of the data. For time-series forecasting, it requires the data to be stationary. The stationarity of this series can be checked by performing an augmented

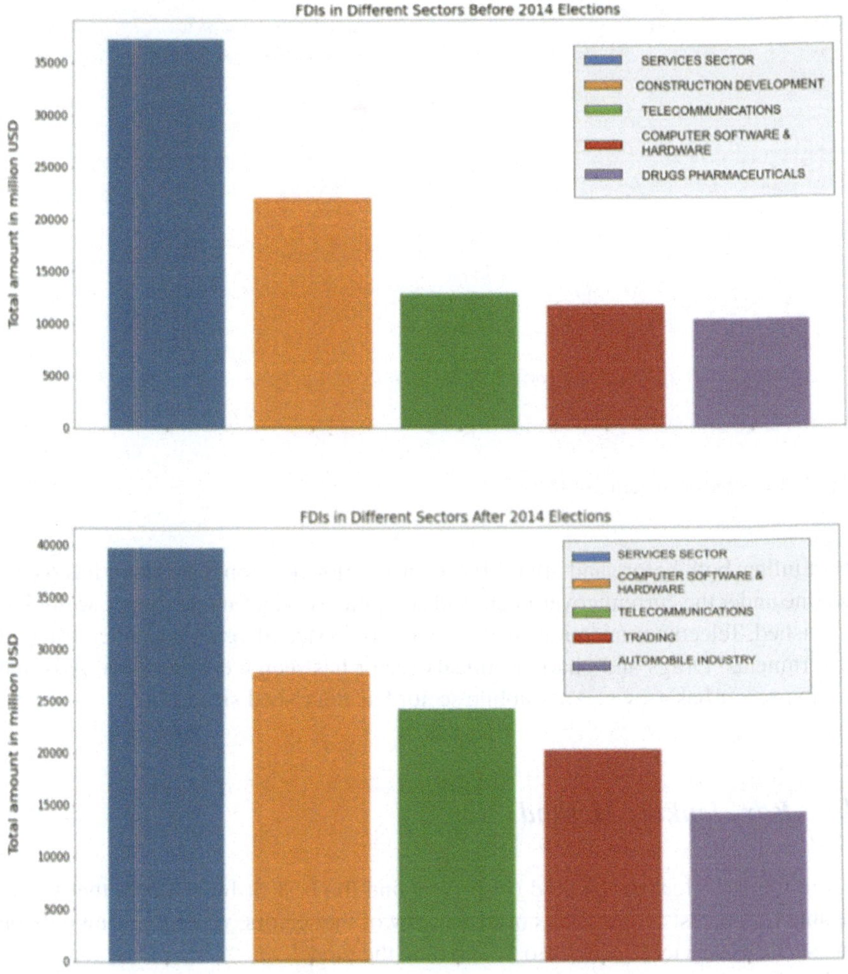

Fig. 6 Top sectors in terms of FDI before (top) and after (bottom) 2014

Dickey–Fuller test on this data, which gives a test statistic of −0.816 and a *p*-value of 0.814 (>0.05). This suggests that the null hypothesis, that the data is non-stationary, is not rejected. Hence, this data is not stationary, and transformation is required.

A non-stationary series can be converted into stationary by applying transformations on it, such as differencing or taking the log of the series. Performing differencing on the data and performing an augmented Dickey–Fuller test on this transformed data gives a test statistic of −3.866 and a *p*-value of 0.002 (<0.05). The null hypothesis is to be rejected, thus this transformed data is now stationary. This can also be verified by plotting the series (Fig. 9). The data is thus stationary and is now suitable to be further processed on.

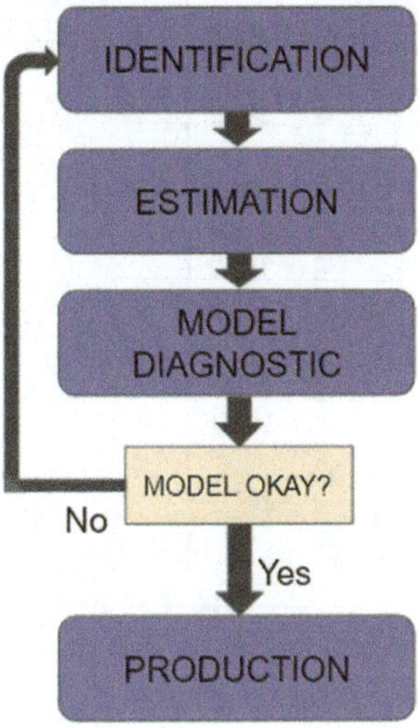

Fig. 7 Box–Jenkins method

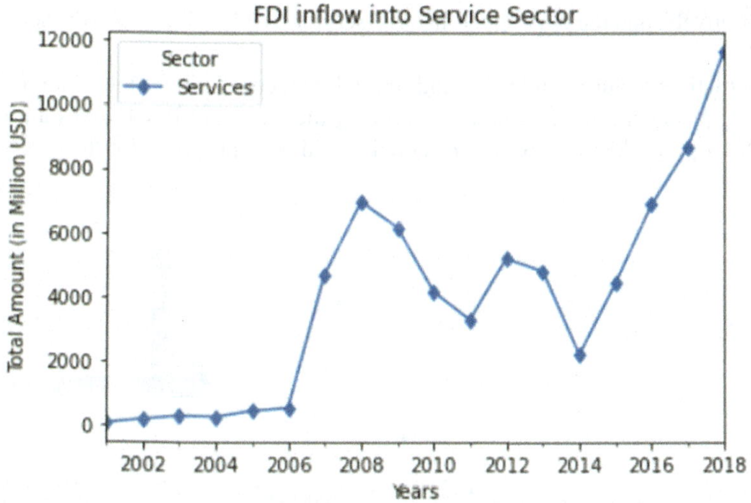

Fig. 8 Plot of the FDI in "Services" sector with time

4.1.2 Estimation

Training the model coefficients using the data is achieved by fitting the data to the model. Trying out different values for p and q to the fitted model and comparing their Akaike information criterion (AIC) and Bayesian information criterion (BIC) help narrow down to more promising values for p and q of the model. The values of p and q are chosen such that the fitted model gives the least values of AIC and BIC. In this case, the value of p and q such that the AIC and BIC is lowest was found to be 0 and

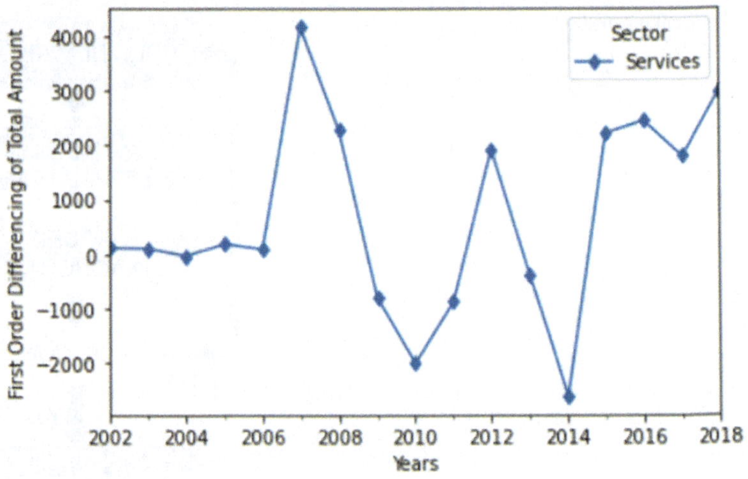

Fig. 9 Plot of the series post transformation

2, respectively, and the corresponding AIC score was 329.53 and the BIC score was 333.09.

Alternatively, autocorrelation and partial autocorrelation plots also help in selecting promising model orders. Analyzing the ACF and PACF gives an idea if an ARMA(p, q), AR(p) or a MA(q) model would be appropriate for the given data.

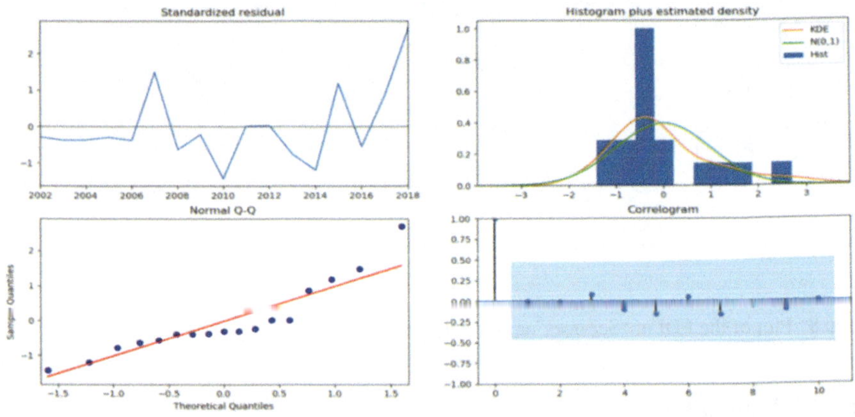

Fig. 10 Model diagnostic plots

4.1.3 Model Diagnostics

Figure 10 gives the different model diagnostic plots as evaluating the quality of the best fitting model is necessary. This can be achieved by examining the test statistics and the diagnostic plots to check the trends of the residuals. A good model should:

- Have no obvious structures in the residuals in the standardized residual plot.
- The KDE and N (0,1) lines in the histogram plus estimated density should almost be the same.
- Have dots lying along the red line in the Q–Q plot.
- 95% of the correlations for lag greater than 0 should not be significant in the correlogram.

4.2 Predicting Future Trends

Forecasting is predicting the future based on current and past trends. It requires a time-series data. Forecasting is very common in the fields of stock market, housing market, and other variable of interest.

Generally, forecasting is used to predict the value at the immediate next time step, using a one step ahead prediction on this series, where output from previous time step is used to predict the value for the consequent next time step, and also focusing only on the "Services" sector gives the following plot (Fig. 11). Data of the first nine years has been used for predicting the values for the consequent ten years.

The light-red shaded area is the confidence interval or the lower and upper limits of the predictions by the model. It captures the trend in the data particularly well, but not very accurately.

The above-fitted model can be used to get an idea about the trend in FDI that can be expected in the services sector in the coming years.

Figure 12 is the forecast that this model has made for the trends that it expects the "Services" sector to follow in the next 5 years.

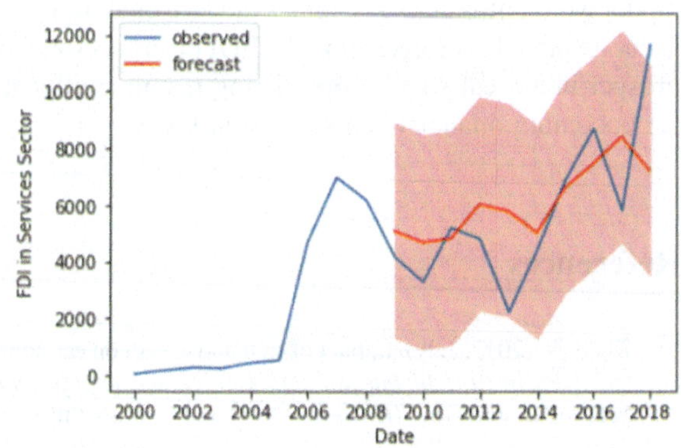

Fig. 11 One step ahead prediction

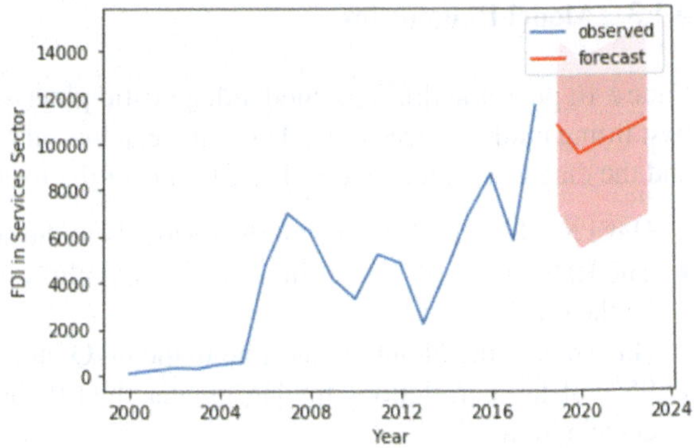

Fig. 12 Predicted trend for the next 5 years

5 Conclusion

The proposed model is able to predict future trends in FDIs in any sector with a certain degree of confidence. Although a huge amount of data will be required to achieve high accuracy, the model requires a thorough hyper-parameter tuning before being deployed. Based on the plots above, it can be seen that the model is able to capture the trends (uptrend or downtrend) in the data.

Such a model can be used to predict various other factors such as unemployment rates in the country, import rates, export rates, stock market price or the country's GDP as a whole, and so on. These are very important factors for finding out any country's growth and decides whether a country is progressing ahead. Predicting such factors gives a fair estimate or idea of where the country is headed and necessary steps can be taken based on the predictions to revive the economy or to prevent any unprecedented downfalls in the economy. To become a super power or be among the top economies in the world, analyzing the present conditions in the economy and using the power of technology to forecast the future trends are very essential.

In the future, we would like to have access to more data and perhaps hope to collect quarterly or monthly, or even weekly data for each sector. ARIMA model fails to capture the seasonal aspects in the data. Using a SARIMA (Seasonal ARIMA) model would also help us capture the seasonal trends so that we can work on a more accurate model for forecasting the FDI in any sector. Doing so would help the country prosper in a great way by optimizing resource allocation and preparing the budget for upcoming financial years in a smart way.

References

1. Khan, N. (2017, July). Impact of FDI and export on economic growth: Evidence from Pakistan and India. In *2017 4th International Conference on Industrial Economics System and Industrial Security Engineering (IEIS)* (pp. 1–5). IEEE. https://ieeexplore.ieee.org/document/8078619.

2. Shaobin, W., & Zhemin, W. (2010, November). Analysis on relationship between FDI and economic growth in ShaanXi based on OLS model. In *2010 3rd International Conference on Information Management, Innovation Management and Industrial Engineering* (Vol. 2, pp. 194–196). IEEE. https://ieeexplore.ieee.org/document/5694550
3. Zhu, J., Wang, H., Xiong, B., & Zhou, F. (2011, May). The impact of FDI on China's economy security: An empirical analysis based on factor analysis & GRNN neural network. In *2011 International Conference on Business Management and Electronic Information* (Vol. 5, pp. 399–402). IEEE. https://ieeexplore.ieee.org/document/5914504.
4. Chen, Y. (2014, September). The analysis of the technical progress effect of FDI in China. In *2014 IEEE Workshop on Advanced Research and Technology in Industry Applications (WARTIA)* (pp. 434–436). IEEE. https://ieeexplore.ieee.org/document/6976288.
5. Guan, J., & Chang, H. (2010, October). The trade effect analysis of foreign direct investment in Hebei. In *2010 International Conference on Future Information Technology and Management Engineering* (Vol. 3, pp. 399–402). IEEE. https://ieeexplore.ieee.org/document/5655683.
6. Li-wei, C., & He, L. (2006, October). Empirical investigation of FDI impact on China's employment. In *2006 International Conference on Management Science and Engineering* (pp. 1123–1127). IEEE. https://ieeexplore.ieee.org/document/4105063.
7. Chao, Y., & Ye, Z. (2011, October). Effects of China's outward FDI on domestic employment. In *2011 10th International Symposium on Distributed Computing and Applications to Business, Engineering and Science* (pp. 137–141). IEEE. https://ieeexplore.ieee.org/document/6118585.
8. Aruna, V., & Gafoor, A. A. (2012, March). Foreign direct investment (FDI) in the muli-brand retail sector in India—A boon or bane. In *IEEE-International Conference on Advances In Engineering, Science And Management (ICAESM-2012)* (pp. 32–35). IEEE. https://ieeexplore.ieee.org/document/6216218
9. Seguino, S. (2000). Accounting for gender in Asian economic growth. *Feminist Economics, 6*(3), 27–58. https://www.tandfonline.com/doi/abs/10.1080/135457000750020128
10. Zhang, K. H. (2001). Does foreign direct investment promote economic growth? Evidence from East Asia and Latin America. *Contemporary Economic Policy, 19*(2), 175–185. https://doi.org/10.1111/j.1465-7287.2001.tb00059.x.
11. Blonigen, B. A., & Piger, J. (2014). Determinants of foreign direct investment. *Canadian Journal of Economics/Revue Canadienne D'économique, 47*(3), 775–812. https://doi.org/10.1111/caje.12091.
12. Stoddard, O., & Noy, I. (2015). Fire-sale FDI? The impact of financial crises on foreign direct investment. *Review of Development Economics, 19*(2), 387–399. https://doi.org/10.1111/rode.12149.
13. Athreye, S., & Kapur, S. (2001). Private foreign investment in India: Pain or panacea? *World Economy, 24*(3), 399–424. https://doi.org/10.1111/1467-9701.00362.
14. Yao, S. (2006). On economic growth, FDI and exports in China. *Applied Economics, 38*(3), 339–351. https://doi.org/10.1080/00036840500368730.
15. Chakraborty, C., & Basu, P. (2002). Foreign direct investment and growth in India: A cointegration approach. *Applied Economics, 34*(9), 1061–1073. https://doi.org/10.1080/00036840110074079.
16. Nandi, S., & Kumar, S. (2005). Variability of exports and imports in Indian perspective: An empirical study. *Asia Pacific Business Review, 1*(1), 68–75. https://doi.org/10.1177/097324700500100108.
17. Shan, J. (2002). A VAR approach to the economics of FDI in China. *Applied Economics, 34*(7), 885–893. https://doi.org/10.1080/00036840110058941.
18. Ciruelos, A., & Wang, M. (2005). International technology diffusion: Effects of trade and FDI. *Atlantic Economic Journal, 33*(4), 437–449. https://link.springer.com/article/10.1007/s11293-005-2871-1.
19. Cheung, K. Y., & Ping, L. (2004). Spillover effects of FDI on innovation in China: evidence from the provincial data. *China Economic Review, 15*(1), 25–44. https://www.sciencedirect.com/science/article/pii/S1043951X03000270

20. Dees, S. (1998). Foreign direct investment in China: Determinants and effects. *Economics of Planning, 31*(2–3), 175–194. https://econpapers.repec.org/article/kapecopln/v_3a31_3ay_3a1998_3ai_3a2_3ap_3a175-194.htm.
21. Cao, X.-y., Analysis on foreign direct investment's effect: The employment of labor force in China. https://en.cnki.com.cn/Article_en/CJFDTotal-TJJY200305008.htm.
22. Granger, C. W. (1969). Investigating causal relations by econometric models and cross-spectral methods. *Econometrica: Journal of the Econometric Society*, 424–438. https://www.jstor.org/stable/1912791?origin=crossref&seq=1#metadata_info_tab_contents.
23. Granger, C. W., & Weiss, A. A. (1983). *Time series analysis of error-correction models. In Studies in econometrics, time series, and multivariate statistics* (pp. 255–278). Academic Press. https://www.sciencedirect.com/science/article/pii/B9780123987501500188.
24. Arellano, M., & Bover, O. (1995). Another look at the instrumental variable estimation of error-components models. *Journal of Econometrics, 68*(1), 29–51. https://doi.org/10.1016/0304-4076(94)01642-D.
25. Kokko, A. (2006). *The home country effects of FDI in developed economies* (Vol. 225). Stockholm: European Institute of Japanese Studies. https://pdfs.semanticscholar.org/a2f9/494724daa90f5f6c222efe133b9551476780.pdf.

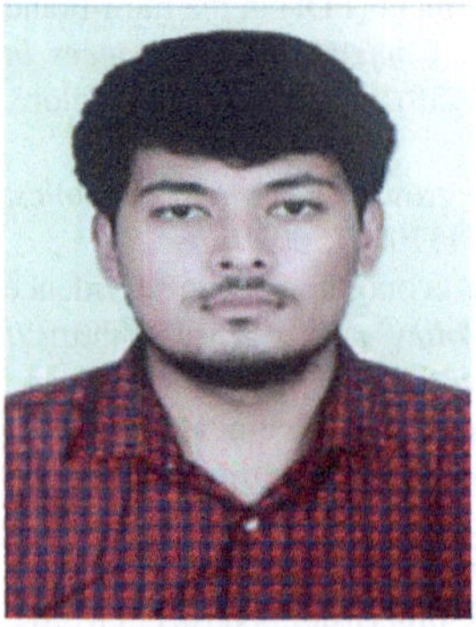

Parikshit Barua is an undergraduate student, currently pursuing B.Tech in Computer Science and Engineering at KIIT Deemed to be University. His area of interest are machine learning, data analytics, finance. He can be reached at parikshitb04@gmail.com.

Sandeepan Mohapatra is an undergraduate student, currently pursuing B.Tech in Computer Science and Engineering at KIIT Deemed to be University. His area of interest are machine learning, data analytics. He can be reached at mahapatrasandeepan@gmail.com.

Siddharth Swarup Rautaray, Ph.D. (Computer Science), Member of IEEE is Professor at the School of Computer Engineering, KIIT University, Bhubaneswar. He has more than a decade of teaching and research experience. Dr. Rautray has published numbers of Research Papers in peer-reviewed International Journals and conferences. His areas of interest is Image Processing/DA/Human Computer Interaction. He can be reached at siddharthfcs@kiit.ac.in.

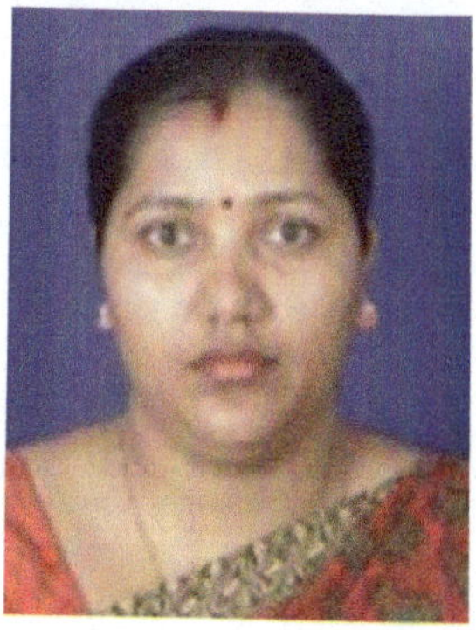

Manjusha Pandey, Ph.D. (Computer Science), is Professor at the School of Computer Engineering, KIIT University, Bhubaneswar. She has more than a decade of teaching and research experience. Dr. Pandey has published numbers of Research Papers in peer reviewed International Journals and conferences. Her areas of interest is WSN, Data analytics etc. She can be reached at manjushafcs@kiit.ac.in.

Customer Profiling and Retention Using Recommendation System and Factor Identification to Predict Customer Churn in Telecom Industry

Nooria Karimi, Adyasha Dash, Siddharth Swarup Rautaray, and Manjusha Pandey

Abstract In the telecommunication area, an immense bulk of data is being created consistently because of a huge customer base. Chiefs and business analyst stressed that accomplishing new customer is prohibitive rather than retaining the current ones. Business experts and CRM analyzers need to know the explanations behind customer's attrition, just as, behavior patterns from the current churn customers so that the business applies some retention strategies to retain the customers before they leave the industry. This paper proposes a, customer churn prediction model that utilizes numerous machine learning classification algorithms, like gradient boost, random forest (RF), decision tree (DT), logistic regression (LR), K-neighbors, in which the gradient boost classifier outperformed other classifiers, and, clustering technique has been utilized to break down and recognize the churn customers and give the factors behind the churn customers in the telecommunication sector. The proposed model initially classifies churn customers data, utilizing classification algorithms, After classification, the proposed model sections the churn customers data by categorizing the customers in groups utilizing k-means clustering algorithm to give group-based retention offers. This paper additionally distinguished churn factors that are basic in deciding the underlying drivers of churn in which the "attribute selected classifier" algorithm has been utilized from Weka tool; by realizing the critical churn factors from customer's data, CRM can improve profitability, prescribe pertinent advancements and relevant promotions to the group of likely churn customers, dependent on similar behavior patterns, and excessively improve marketing campaigns of the organization (Ullah, Irfan, et al. "A churn prediction model using random forest:

N. Karimi (✉) · A. Dash · S. S. Rautaray · M. Pandey
School of Computer Engineering, KIIT (Deemed to be University), Bhubaneswar, Odisha, India
e-mail: Noria.karimi2@gmail.com

A. Dash
e-mail: adyasha.dashfcs@kiit.ac.in

S. S. Rautaray
e-mail: siddharthfcs@kiit.ac.in

M. Pandey
e-mail: manjushafcs@kiit.ac.in

M. Pandey and S. S. Rautaray (eds.), *Machine Learning: Theoretical Foundations and Practical Applications*, Studies in Big Data 87,
https://doi.org/10.1007/978-981-33-6518-6_9

analysis of machine learning techniques for churn prediction and factor identification in telecom sector." EEE Access 7 (2019): 60, 134–60, 149). In addition, recommendation system has been utilized for customer retention. As creating retention strategies is a basic undertaking of the CRM to prevent customers from attrition. The proposed churn prediction model is assessed utilizing measurement matrices, for example, accuracy, precision, recall, f-measure, and receiving operating characteristics (ROC) area. The outcomes uncover that the proposed customer churn prediction model delivered better churn classification utilizing the gradient boost algorithm and customer profiling, utilizing k-means clustering. Moreover, it additionally gives factors behind customer's turnover, through the rules produced by utilizing the attribute selected classifier algorithm. In the end of the examination, some retention strategy has been applied using recommendation system to maintain and keep up the customers for long time base.

Keywords Churn prediction · Telecom · Machine learning · Factor identification · Customer profiling · k-means clustering · Recommendation system

1 Introduction

In the present world, telecom organizations at an exceedingly quick rate are producing an enormous volume of data. There is a scope of telecom specialist organizations contending in the market to build their customer base. Customers have numerous alternatives as better and more affordable services. A definitive objective of telecom organizations is to boost their benefit and remain alive in a focused commercial center [1]. A customer attrition happens when a huge level of customers are most certainly not happy with the administrations of any telecom organization; customer churn additionally influences the general notoriety and reputation of an organization, which brings about its image misfortune and brand loss. The contender organizations occasionally influence a faithful customer, who produces high income for the organization. Such customers expand the benefit of an organization by alluding it to their companions, relatives, and associates [2].

Telecom organizations consider strategy move when the quantity of customers dips under a specific level, which may bring about an enormous loss of income [3]. Churn prediction is imperative in the telecom division as telecom operators need to hold their important customers and improve their customer relationship management (CRM) administration [4, 5]. The most testing and challenging activity for CRM is to hold and retain existing customers [6]. Because of the soaked and aggressive market, customers have the alternative to change to other specialist organizations.

Telecom organizations have created methods to recognize and maintain their customers, as it is more affordable than pulling in the new ones [4]. This is because of the cost engaged with commercials, workforce, and concessions which can scale up to just about five to multiple times than holding and maintaining existing customers [3]. Little consideration is required for recognizing the current churn customers, which can help in upsetting the circumstance. The necessity of holding customers needs to build up an exact and elite model for recognizing churn customers.

As well as the proposed model ought to have the capacity to distinguish churn customers and afterward discover the explanations for customer churn to keep away from loss of customers and give measures to hold them. What is more, it should utilize methods to anticipate when such a circumstance will emerge later on. Because of ongoing headway in the field of big data, there exist numerous data mining and machine learning solutions, which can be utilized to analyze and break down such data. These methods break down the data and distinguishes reasons for customer churn; CRM can utilize these systems to augment their benefit [2].

It is difficult to anticipate the reason for churn and its recurrence. Various issues with customer advancement emerge basically in light of the quality component including service quality, network coverage, load errors, billing, costs, technologies, innovations, and so on [2]. These service quality elements enable customers to contrast service quality and benefits and other good service suppliers [4]. The forecast pace of the ordinary customer in the telecom division is assessed at 2%, which is all out yearly loss of roughly 100 billion dollars [7].

Existing examinations uncover that the essential goal is to utilize a huge volume of telecom data to distinguish the significant churn customers [2]. An enormous volume of data is being produced in the telecom area, and the data contains missing qualities, which lead to the poor consequences of the expectation models. To deal with these issues, data pre-processing techniques are adjusted to expel clamor from data, which is powerful for a model to accurately group the data and improve the exhibition. Feature selection has been utilized in literature; however, various data feature selection are dismissed while model development [8].

In this examination, the proposed model for customer attrition has been created that utilizes different machine learning algorithms. The accuracy of a classifier relies upon the accessible data set [2]. It is approved by utilizing a telecom data set, which is publically accessible on Kaggle web site. The proposed customer churn prediction model is assessed utilizing information retrieval metrics. The accuracy is calculated for churn prediction model utilizing TP rate, FP rate, precision, recall, F-measure, and ROC area; the target of the examination is to research the current procedures in machine learning and data mining to profound a model for customer churn prediction and recognize the factors behind customer churn, and customer profiling has been utilized to distinct customers based on their behavior as it depends on the behavior of customers, so that apply retention strategies for preventing the customers from leaving the industry. The gradient boost classifier algorithm created better accuracy and precision, while contrasted with other machine learning algorithms.

The customer profiling distincted the customers based on their behavior into three groups high, medium, and low, utilizing k-means clustering algorithm. The factors behind customer churn has been found by utilizing the rules created from "attribute selected classifier" with the help of Weka statistical tool.

Rest of the paper is arranged as follows. Section 2 gives literature survey. Section 3 defines the business problem, and Sect. 4 shows the proposed model for customer churn prediction. Section 5 discusses the factors identifying customer churn, and

Sect. 6 gives customer profiling and retention guidelines. Section 7 discusses the customer retention using recommendation system, and Sect. 8 depict results and experiments. Sect. 9 finishes up the talk and gives future work.

2 Literature Review

In this chapter, a review of the bibliographic content found is conducted. The focus is the methodologies and strategies applied to customer churn prediction and identifying the factors behind customer churn as well as customer behavior investigation to foresee churn in telecom businesses. Through customer profiling in CRM, media transmission has been acted in the literature, utilizing different methods including machine learning, data mining, and big data analytic. These procedures bolster organizations to investigate customer churn, recognize, anticipate, and maintain churn customers, help in basic leadership and CRM. Long ago many research has been done in the customer relationship the executives in media transmission in the method for breaking down the customer's behavior and foreseeing the customer churn, here are a few works that has been done in the referenced territory [8].

Gavril et al. (2016) introduced a propelled philosophy of information digging to anticipate churn for prepaid customers utilizing data set for call subtleties of 3333 clients with 21 features and a reliant churn parameter with two qualities: Yes/No. A few features incorporate data about the quantity of approaching and active messages and phone message for every customer. The creator applied head segment examination calculation "PCA" to diminish information measurements [9].

Three machine learning algorithms were utilized, neural networks, support vector machine, and Bayes networks, to foresee churn factor. The creator utilized AUC to quantify the exhibition of the calculations. The AUC esteems were 99.10, 99.55, and 99.70% for Bayes networks, neural systems, and bolster vector machine, respectively. The data set utilized in this investigation is little, and no missing qualities existed. Another element choice strategy is proposed to determine CRM data index with pertinent highlights by joining a productive information mining methods to improve data quality and feature importance [10].

Huang et al. (2015) contemplated the issue of customer churn in the large data platform. The objective of the scientists was to demonstrate that huge data enormously improve the way toward anticipating the churn contingent upon the volume, assortment, and speed of the data [11].

Makhtar et al. (2017) proposed a model for churn prediction utilizing rough set theory in telecom. As referenced in this paper, rough set characterization calculation outflanked different calculations like linear regression, decision tree, and voted perception neural network. Different investigates considered the issue of lopsided informational indexes where the stirred client classes are littler than the dynamic client classes, as it is a significant issue in beat expectation issue [12].

Amin et al. (2013) looked at six different inspecting strategies for oversampling with respect to telecom churn predicting issue. The outcomes demonstrated that the calculations (MTDF and rules age dependent on hereditary calculations) outflanked the other analyzed oversampling calculations [13].

Amin et al. (2018) The proposed study not only presented a novel JIT-CCP model for the telecom company where: (i) a company is newly established and do not have historical data to build an efficient CCP model, (ii) a company has lost the data due to any special reason, or (iii) recently adopted the modern technology for maintaining the customer behavior, but also demonstrated the comparison of the effect of with and without data transformation methods (e.g., log and rank) for the JIT-CCP [14].

Ullah et al. (2019) this research aimed to build a system that predicts the churn of customers in SyriaTel telecom company. Four tree-based algorithms were chosen because of their diversity and applicability in this type of prediction. These algorithms are decision tree, random forest, GBM tree algorithm, and XGBOOST algorithm. The method of preparation and selection of features and entering the mobile social network features had the biggest impact on the success of this model, since the value of AUC in SyriaTel reached 93.301%. XGBOOST tree model achieved the best results in all measurements [2].

In paper [15], they recognized significant customers, anticipated future behavior, and empower firms to make proactive, information-driven choices by data mining strategies. So principally, it centers around dissecting shopper data for monetary advantages.

In paper [16], they characterize the chose customer into bunches utilizing RFM model to distinguish high benefit, gold clients. They utilize data mining devices and successfully find the present spending example of customers and patterns of social change, which will enable administration to recognize in an enormous database potential changes of customer inclinations and give as ahead of schedule as potential items and administrations wanted by the clients to extend the customer base and forestall client steady loss.

Jegan and Sudalaiyandi (2012) directed an examination study on consumer behavior toward cell phone benefits in Kovilpatti, Thoothukudi area on 100 portable clients and reasoned that the components that impact the purchasers to purchase a specific cell phone administrator is called call duties followed by organize inclusion and brand picture [17].

Huang et al. [11] contemplated the issue of customer churn in the large information platform. The objective of the scientists was to demonstrate that huge information enormously improve the way toward anticipating the beat contingent upon the volume, assortment, and speed of the information. Managing information from the operation support office and business support division at China's biggest broadcast communications organization required a major information stage to design the breaks. Arbitrary forest calculation was utilized and assessed utilizing AUC.

The novel model introduced in [8] shows a hybrid approach connecting the adjusted k-means grouping calculation with the old style rule inductive system (FOIL) to foresee churn customer behavior. The management of a large volume of data in the present world provides an opportunity to improve the quality of administration to the customers. This data includes data about customers conduct, use example and system operations. In this paper, in the main stage, customers are sectioned

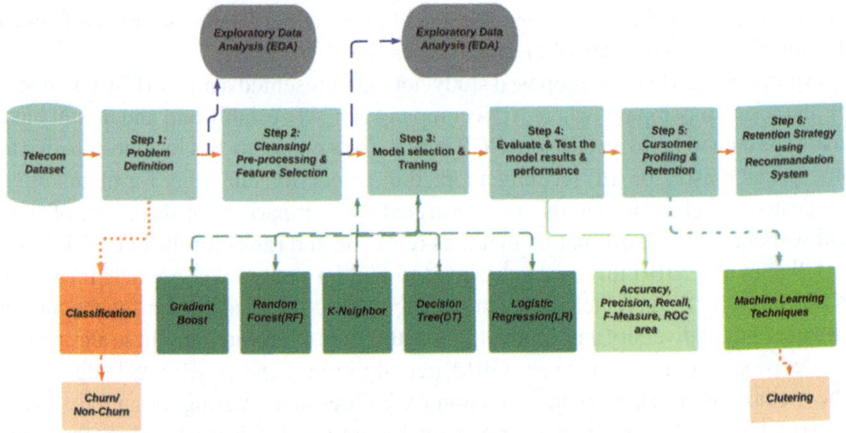

Fig. 1 Proposed model for customer churn prediction

utilizing choice standards, and in the subsequent stage, a model is created for each leaf of the tree. This half breed approach is contrasted and calculated relapse, choice trees, arbitrary woodlands, and strategic model trees [18] for churn prediction.

In this examination, the feature-engineering phase is mulled over to make our very own features to be utilized in machine learning algorithms. In this examination, gradient boost classifier, support vector machine, random forest, N-nearest neighbor, utilized and strategic relapse in which the gradient boost classifier produced the high accuracy and precision which has been utilized for customer churn prediction, for identifying customer churn factors the Attribute selected classifier from weka tool has been utilized, as well as k-means clustering algorithm has been used for customer profiling to dissect customers based on their similar behaviors patterns, and also recommendation system has been performed for customer retention in telecom sector. The customer churn prediction model is shown in Fig. 1.

3 Problem Definition

This paper proposes a model as shown in Fig. 1: The problem has been identified into the following four phases.

Phase 1: Customer churn identification

Customer churn, also known as customer attrition, in its most basic form, is when a customer chooses to stop using your products or services [19]. To start understanding the customer's journey and the experiences they have, there should be a relevant and holistic model to tackle this issue. For this, machine learning algorithms and data mining techniques have been utilized, and a model has been built for customer churn prediction, which has been shown in Fig. 1.

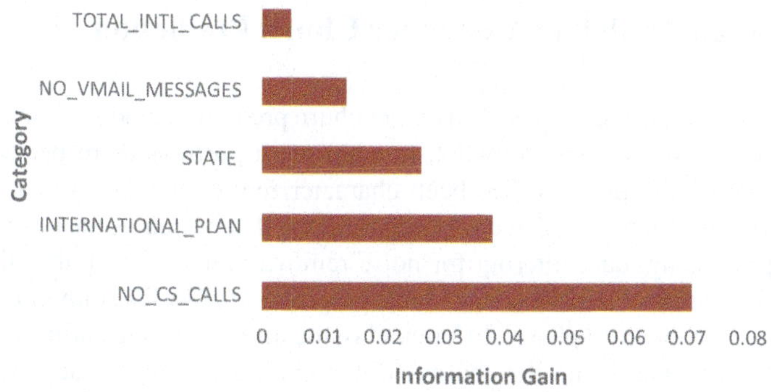

Fig. 2 Chart representation of Table 3

Phase 2: Identifying factors responsible for customer churn

The factors of churn customers are examined in this subsection, which is classified by the attribute selected classier algorithm. This technique gives rules for churn prediction and gives customer churn behaviors and patterns [2]. These key principles are truly important for the leaders to maintain the customers who are likely to leave the company. The attribute selected classifier algorithm characters numerous reasons of churn and gives features which rely upon one another. The churn-related principles given by the attribute selected classier algorithm is shown in Table 3 and Fig. 2.

Phase 3: Customer profiling and retention

A gigantic volume of information is being created every day because of a tremendous customer base; beside customer migration has identified big concern for telecom companies. Business examiners and customer relationship management (CRM) analyzers need to know behavior patterns of customers, from the existing customer's data [3]. To recognize loyal customers and unfaithful customers, based on that, take some business strategies, to maintain them for long-time relationship. The proposed model categories the customers based on their behavior in the telecom company and recognizes the churn and non-churn customers.

Phase 4: Customer retention using recommendation system

For decreasing customer attrition, there ought to be some business strategy the business administers and decision makers should apply to not solely attract customers, however, retain and hold them for long-time amount. This is often what this paper is all regarding. This experiment develops a recommendation system to assist telecom business to retain their customers. The model is shown in Fig. 1.

4 Proposed Model for Customer Churn Prediction

This chapter exhibits the proposed customer churn prediction model. Figure 1: shows the proposed churn prediction model. In addition, it portrays its respected phases. In initial phase, the problem has been characterized, in the subsequent advance, exploratory data analysis utilized for preparing the data for additional examination, data pre-processing, data filtering for noise removal, removal of imbalanced data, and standardization of data has done in this progression as well as important features are extracted, in the third phase, different classification algorithms utilized, for categorizing the customers into the churn and non-churn customers. The classification algorithms incorporate gradient boost, random forest (RF), K-neighbor, decision tree (DT), logistic regression (LR), and in the fourth phase, the assessment and testing of the model is performed utilizing confusion matrix, Precision and Recall, F-measure and ROC area; in the fifth step, customer profiling is performed utilized k-means clustering technique. In the final step, the model recommends retention strategies for each category of churn customers. In this investigation, a range of machine learning algorithms are utilized for classifying customer's data using the labeled data set. It is to evaluate which of the algorithm best classifies the customers into the churn and non-churn categories. In the first place, the gradient boost algorithm is utilized for classification. Furthermore, random forest, decision tree, K-nearest, and logistic regression utilized, with tenfold cross-validation. The classification process has performed by machine learning algorithms, and Weka tool has been utilized for factor identification.

4.1 Data Set Description

In this investigation, two data sets have been utilized. The first data set is used for training the model and the subsequent one is utilized for testing the model, this data set has 3333 instances with 21 features, in which exploratory data analysis and feature selection method has performed, with 13 relevant and useful features have produced for training and testing the model, as well as feature engineering has done utilizing machine learning libraries, it contains labeled data with two classes in which 14% of the complete data is named as churn "YES" though 85.5% perception have the value "NO" non-churn. The data set variables of customer transactions and their descriptions are introduced in Table 1, The data sets are accessible on Kaggle site https://www.kaggle.com/c/churn-analytics-bda/data which are both in numerical and categorical structure; the data exploratory analysis has been utilized in the subsection to defeat the issues with the data sets, and for validating the data, the test data set has been utilized. Table 1 describes the data set with the related data types.

Table 1 Customer data set description

#S	Variable	Data type	Description
1	AccountLength	Integer	How long account has been active
2	Int'l Plan	Categorical	International plan activated (Yes, No)
3	Vmail Plan	Categorical	Voice Mail plan activated (Yes, No)
4	VamilMessage	Integer	No. of voicemail messages
5	DayMins	Integer	Total day minutes
6	DayCalls	Integer	Total day calls made
7	DayCharge	Integer	Total day charge
8	EveMins	Integer	Total evening minutes
9	EveCalls	Integer	Total evening calls
10	EveCharge	Integer	Total evening charges
11	NightMins	Integer	Total night minutes
12	NightCalls	Integer	Total night calls
13	NightCharge	Integer	Total night charges
14	IntlMins	Integer	Total international minutes used
15	IntlCalls	Integer	Total international calls made
16	IntlCharge	Integer	Total international charges
17	CustServCalls	integer	Number of customer service calls made
18	Churn	Categorical	Customerchurn (Yes = churn, No = no-churn)

4.2 *Data Pre-processing*

Data pre-processing is a significant advance on any data mining venture. Real-world data is commonly deficient, conflicting and even blunders. To create a decent model to an issue, these issues must be addressed. The problems with the data set are listed in the sub-section below.

a. **Missing data**: In this examination, we have dealt with this issue by utilizing mean and mode for supplanting the missing values.
b. **Noise Removal**: There are many missing values in telecom data set, wrong qualities like "invalid" and awkwardness characteristics in the data set. In this data set, the amount of features is 21. The filtering technique has been done for separating the unnecessary data form relevant data; the amount of features after applying filtering technique is show in Table 2.
c. **Redundant data**: This sort of error is for the most part related with human blunders. The data could have been recorded under various names or in different places. For our situation, fortunately there are no redundant data.
d. **Unbalanced Data set**: The created data set was imbalance since it is an uncommon instance of the classification issue where the appropriation of a class is not normally homogeneous with other classes. The predominant class is known as the essential class, and the others are known as the optional class.

Table 2 Relevant features

Attributes	Important features ranking
No_CS_Calls	0.20875
International_Plan	0.25985
State	0.01874
Voice_Mail_Plan	0.10215
Total_Intl_Calls	0.05284
Phone_No	0.00573
Total_Night_Calls	0.00614
Account_Length	0.01654
Area_Code	0.00617
Total_Night_Charge	0.0355
Total_Night_Minutes	0.033549
Total_Day_Calls	0.01846

Table 3 Ranked gain information of factors in various categories

S#	Category	Information Gain
1	NO_CS_CALLS	0.07072
2	INTERNATIONAL_PLAN	0.03787
3	STATE	0.026616
4	NO_VMAIL_MESSAGES	0.0138
5	TOTAL_INIL_CALLS	0.00467

Data set is unbalanced if one of its categories is 10% or less contrasted with the other one [7]. Data set was unbalanced since the level of the optional class that speaks churn customers is about 14.5% of the entire data set, and the targeted class is 85%. It is clearly shown that the data is skewed. Along these lines, if the machine ge, with this data set the machine will produce the outcome as true or churn with; accordingly, we utilize a few strategies for taking care of this issue. There are numerous techniques for taking care of the imbalance issue, which are under-sampling, over-sampling, and Smote; each one is utilized for dealing with imbalance data. In this experiment, we have utilized Stratified-K-Fold method for taking care of the issue.

e. **Features Selection**: Feature selection is a critical advance for choosing the important features from a data set dependent on domain knowledge. Various techniques exist in the literature for feature selection [1, 20]. With regard to churn prediction, in this examination, we utilized feature_importance method. Feature importance method gives a score for each feature of the data; the higher the score, the more important or relevant is the feature toward your output variable [21]. In churn data set, we chose just the main 13 relevant features out of the complete 21 features, having high-ranking qualities in the result of two techniques. The performance of classification increments the data set contains profoundly prescient and important factors [2].

Besides, the data sets with high relationships has dropped from both the data sets which are as follows. TOTAL_DAY_MINUTES, TOTAL_DAY_CHARGE, VOICE_MAIL_PLAN, NO_VMAIL_MESSGES, TOTAL_NT1_CHARGE, TOTAL_NT1_MINUTES, TOTAL_EVE_MINUTESTOTA, L_EVE_CHARGE,VOICE_MAIL_PLAN AND NO_VMAIL_MESSAGES, as they do not give us any new information they are only a relationship, the relevant features has shown in Table 2.

4.3 Performance Evaluation Matrix

In this examination, the profound customer churn prediction model is surveyed using accuracy, precision, recall, F-measure, and ROC area. Equation 1 ascertains the accuracy metric. It distinguishes number of instances that were correctly classified.

$$\text{Accuracy} = \frac{\text{TP} + \text{TN}}{\text{TP} + \text{TN} + \text{FP} + \text{FN}} \tag{1}$$

TP rate is generally called sensitivity. It notices the portion of data which is precisely named positive. In addition, it must be high for any classifier. It has depicted by in Eq. 2.

$$\text{TP Rate} = \text{True Positive}/\text{Actual positive} \tag{2}$$

FP rate uncovers which, some part of the data are erroneously assigned positive, and it must be low for any classifier. As t is controlled by utilizing Eq. 3:

$$\text{FP Rate} = \text{False Positive}/\text{Actual Negatives} \tag{3}$$

Accuracy is generally called positive predictive value (PPV), and it shows which part of the predicted data is positive. It is controlled by utilizing Eq. 4.

$$\text{Precision} = (\text{True Positive})/(\text{True Positive} + \text{False positive}) \tag{4}$$

The recall is another measure for completeness instances, the likelihood that all the applicable features are picked by the system. It is controlled by utilizing Eq. 5.

$$\text{Precision} = (\text{True Positive})/(\text{True Positive} + \text{False Negative}) \tag{5}$$

The F-measure value is a trade-off between accurately classifying all the data points and guaranteeing that each class contains points of just one class. It is determined by utilizing Eq. 6.

$$\text{F-measure} = 2 \times (\text{Precision} \times \text{Recall})/(\text{Precision} + \text{Recall}) \quad (6)$$

ROC area demonstrates the normal execution against all possible cost proportion between FP and FN. In the event that the ROC area value is identical to 1.0, this is a perfect prediction. In this manner, the values 0.5, 0.6, 0.7, 0.8, and 0.9 speak to irregular expectation, horrendous, moderate, good, and predominant, respectively.

Values of ROC area other than these show something is not right [2].

5 Factors Identifying Customer Churn

In light of the examination, network quality, call facilities, Internet facilities and booster facilities were the high effect factors that cause customer churn in telecommunication [21]. The objective of this section is to identify the high effective factors that cause customer attrition. The factors of churn customers are examined in this subsection, by utilizing the attribute selected classifier algorithm with info_Gain_Attribute_Evaluator method in Weka tool; this strategy gives rules for customer churn prediction as well as identifying customer conduct patterns; these key principles are entirely significant for the leaders and decision makers to maintain the customers who are more likely to churn. The attribute selected classifier algorithm recognizes numerous reasons behind churn and gives highlights, which rely upon one another [2]. In this experiment, there has been found top five high churn ranking features which cause customer churn and are depicted in Table 3 and in Fig. 2. These five factors are highly dependent features for customers churn, which ought to be considered by decision makers, for retaining them, by taking some vital strategic business planning.

6 Customer Profiling and Retention

Customer cluster is used to fragment the all out customer's data into bunches reliant on their conduct data and their relationship. There are many clustering algorithms which can be applied for different bunching, as various leveled, fuzzy and parcel clustering [2]. This paper grasps k-mean clustering algorithms that are the best for division, which can section the data into different gatherings as the given issue incorporates exceptionally perplexing, heterogeneous, and gigantic data collection [22]. The k-means is a remarkable iterative approach to segment the data. In this technique, the data is divided as having a spot with one of the k-groups. K-means is useful to find a relationship and concealed example in data, which have a place with one class [23]. In this examination, the k-means algorithm parts the data into three groups; elbow methods are used for picking the k-esteem. The three gatherings speak to, high, medium, and low customers. Figure 3 exhibited the three groups with their respected features, and Fig. 4 shows the edge estimation of each bunch. The k-means clustering technique hopes to part the arrangement of in perceptions ($x1$, $x2,\ldots xn$), into $K(\leq \mathrm{i})$ disjoint sets is $= \{S1, S2,\ldots, Sk\}$ to limit the whole of squares inside the bunch [2]. Numerically, this target can be cultivated by using Eq. 7.

```
Cluster 0: WA,143,510,3404989,' no',0,120,104,182.5,85,8.21,4,3
Cluster 1: OH,65,408,3367600,' no',0,116,117,227.3,86,10.23,6,1
Cluster 2: NV,160,415,3333531,' no',0,90,115,263.9,95,11.88,4,1
Final cluster centroids:
```

Attribute	Full Data (3333.0)	Cluster# 0 (859.0)	1 (1308.0)	2 (1166.0)
State	WV	WA	OH	NV
Account_Length	101.399	100.6484	101.0413	102.3533
Area_Code	436.8122	506.0233	412.581	413.006
Phone_No	3755715.7204	3751220.9884	3975251.4832	3512755.3405
International_Plan	no	no	no	no
No_Vmail_Messages	7.8269	6.8999	8.7523	7.4717
Total_Day_Calls	100.6466	99.3306	100.8654	101.3705
Total_Eve_Calls	100.2217	100.0698	102.1705	98.1475
Total_Night_Minutes	200.2605	194.8321	205.1539	198.7702
Total_Night_Calls	99.5284	99.5576	100.6613	98.2358
Total_Night_Charge	9.0118	8.7674	9.2321	8.9447
Total_Intl_Calls	4.4491	4.3923	4.3861	4.5617
No_CS_Calls	1.6415	1.7031	1.7141	1.5146

Fig. 3 Customer clusters and their respected features

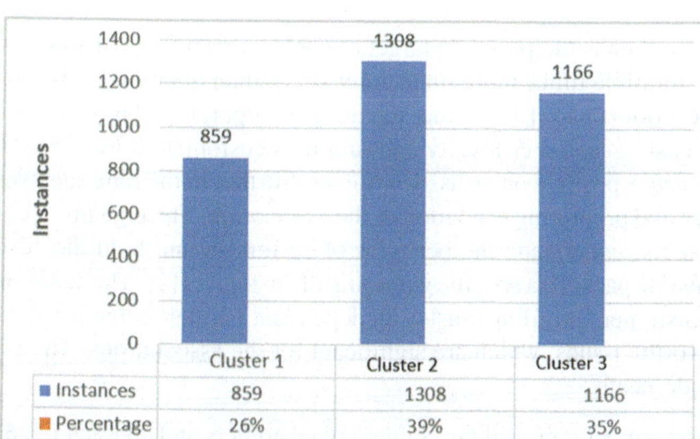

Fig. 4 Segmentation of churn customers

$$j = \min \sum_{i=1}^{k} \sum x \in S1 \, ||x - \mu 1||2 \tag{7}$$

where $\mu 1$ is the mean of point Si. For our circumstance, three groups are made by three distinctive customer conduct. In existing examinations, the researchers investigated the conduct of both classes churner and non-churner [24]. Regardless, our philosophy is extraordinary, as we are looking at the conduct of churner just and find the conduct of comparable customers considering the way that the leaders are intrigued to find the conduct of churner and make fitting strategy for support to help association's advantage. We focused on churn class data just to evaluate the conduct of comparative customers and take a gander at their desire in which the data is allocated into three parts, which join high, medium, and low customers for arrangement making to hold the churn customers. Decision producers and pioneers can undoubtedly comprehend the conduct of a gathering of customers that are dynamically huge and need a real way to deal with improve the maintenance procedure, which is beneficial for the association. For example, Fig. 4 shows that the cluster 2 and cluster 3 have most extreme comparative churn customers 39% and 35%, respectively. These two clusters are increasingly important for the organization to expand the advantage by holding them contrasted with cluster 1 with simply 26% churners. In view of division, we can without much of a stretch locate a comparable example and components of comparative churn customers from this example and conduct; we make rules for the suggestion of just comparable customers later on in the future. These three groups speak to a novel assortment of conduct, and attributes can expand the degree of basic leadership for churner. Utilizing arrangement, chiefs and pioneers can undoubtedly deliver various approaches and customized activities [2] (Fig. 3).

7 Customer Retention Using Recommendation System

Customer retention is the principle target of CRM as retaining existing customers is at least 5 to multiple times more financially savvy than obtaining new ones, relying upon business domains [2]. Customer retention incorporates all moves made by organization to ensure customer loyalty and diminish customer churn [25]. The impact of maintenance is positive on an association profile that in the long run overhauls the maintenance and promoting execution of the association. Through this technique, the organization can understand the behavior of customers and hold the customers by offering sensible packs to a specific gathering of customers [2]. This assessment work shows the customers profiling clustering dependent on their behavior and addresses some noteworthy issues, which are significant for the associations. These issues are depicted underneath.

a. This assessment removed the conduct of customers and characterized them into groups by applying the data mining strategies and focusing on the CRM, to keep and hold customers.

b. The segmentation model distinguished the behavior of different customers as sections, to retain them in groups—high, medium, and low, on a long haul premise.
c. From results and customer practices and behaviors, we can moreover achieve "coordinated" advertising which underscores promoting with particular customers.
d. To satisfy the necessities of customers, their one of a kind conduct and inclinations are perceived to offer them a combination of bundles and services.
e. CRM needs to amass a specific association with customers thinking about their perceived needs and conduct to fabricate a strong association with them.
f. CRM framework should be consolidated among customers and operational bleeding edge framework to effectively administer customers needs.

For holding customers, displaying a planned framework could be effective over the channel for offering different services, for instance, gainful and proficiency services, altered services, and value control services. In this particular condition, CRM is stressed over modifying the association with the customers and association response to the consumer loyalty's, for instance, developing a common correspondence framework, extending front working environments to join all delegates, accomplices, and suppliers to speak with customers through email, telephone, site pages, contacts, content, etc. Through segmentation, the customer's behavior is checked and followed to recognize the patterns and usage, a quick promoting system is a holding technique through which customers are held and a combination of services are offered through various stations [2].

Direct methods are dynamically practical which can decrease the cost and addition effectiveness by offering different services and modified offers directly [3]. In the immediate correspondence with customers, the association shows signs of improvement reaction and can rapidly take a choice from the customer's response; the association can moreover develop a model that can assess likelihood of churn, perceive comparable gathering of customers subject to their reactions, and is progressively useful and valuable for chiefs and decision makers [3].

We pick the similarity-based way to deal with make redid maintenance exercises. This procedure is the most notable instance of a recommender system. Recommender system, using a comparability measure for focusing on comparable customers, depends on the going with methodologies [2].

Content-based: It relies upon past behavior and prescribes comparative inclination and things to a similar customer.

Collaborative: It relies upon past behavior and inclination only for those customers who have a similar inclination and comparative behavior.

Hybrid: It is the mix of both content and collaborative methodology.

This model relies upon the collaborative procedure for customers who have similar past behavior and similar inclinations. Each customer is classified in the high, medium, and low classes to complete altered maintenance techniques later on in the future. Subsequently, the customers are characterized into groups by using rules and an edge esteem which is extricated from clusters. The inside thought of this prescribed framework is to recognize similar churn customers with respect to similar behavior and pattern. Around the completion of the gathering, leaders and boss offer the churn customers maintenance bundles as per the predefined groups.

8 Implementation and Results

This experiment has performed various examinations on the proposed model for identifying the customers who are likely to churn and also segmenting customers dependent on their behavior information and relationship utilizing k-means clustering algorithm. These algorithm clusters customers based on their similar behavior information to distinguish the customer who are more likely to leave the company for another service provider. In this experiment, first the proposed model identifies the churn customers from the available telecom data set; after that the factors that makes the customers to churn has identified which is crucial for telecom industry to know the reasons for customer's turn over; after distinguishing the factors, the customers has been distinguished and clustered based on their similar behaviors, by utilizing k-means clustering algorithm which segmented the customer into three different clusters high, medium, and low customers, as it is shown in Fig. 4. The three clusters present different customers that show the customer's situation regarding the telecom company, and after the segmentation in the final phase of the experiment, the recommendation system has developed to recommend retention strategies for each and every customers to retain and keep the customers for long-time base.

9 Conclusion and Future Work

In the present serious market of telecom area, churn prediction is an enormous issue of the CRM to hold noteworthy customers by recognizing comparable gatherings of customers and giving centered offers/services to the specific gatherings. Thus, here, the scientists have been looking key factors of churn customers to hold them and deal with the issues of CRM and decision maker of the association [2]. In this manner, a customer churn model is obliged for data analytic and affirmed through standard evaluation matrices. The results show that the proposed customer churn prediction model performed better by using machine learning algorithms. Gradient boost, random forest (RF), support vector machine (SVM), logistic regression, K-nearest and decision tree (DT), in which the gradient boost classifier beat the others with 96% accuracy and 85% precision. The fundamental churn factors perceived and performed customer profiling according to their danger of churn. At long last, we gave rules on customer maintenance for chiefs and decision makers of the telecom associations.

In future, we will utilize social network analysis (SNA) to extract statistical and SNA highlights for telecom customers, which enhance the consequences of predicting the churn in telecommunication. And the examination can be additionally reached out to investigate the changing behavior patterns of churn customers utilizing other machine learning and AI techniques, as well as expanding and developing recommendation systems for customer loyalty.

References

1. Babu, S., Ananthanarayanan, N. R., & Ramesh, V. (2014). A survey on factors impacting churn in telecommunication using datamininig techniques. *International Journal of Engineering Research & Technology (IJERT), 3*(3).
2. Ullah, I., et al. (2019). A churn prediction model using random forest: Analysis of machine learning techniques for churn prediction and factor identification in telecom sector. *IEEE Access, 7*, 60134–60149.
3. Verbeke, W., et al. (2011). Building comprehensible customer churn prediction models with advanced rule induction techniques. *Expert Systems with Applications, 38*(3), 2354–2364.
4. Idris, A., & Khan, A. (2012). Customer churn prediction for telecommunication: Employing various various features selection techniques and tree based ensemble classifiers. In *2012 15th International Multitopic Conference (INMIC)*. IEEE.
5. Kaur, M., Singh, K., & Sharma, N. (2013). Data mining as a tool to predict the Churn Behaviour among Indian bank customers. *International Journal on Recent and Innovation Trends in Computing and Communication, 1*(9), 720–725.
6. Miguéis, V. L., et al. (2012). Modeling partial customer churn: On the value of first product-category purchase sequences. *Expert Systems with Applications, 39*(12), 11250–11256.
7. Manzano-Machob, D. (2013). The architecture of a churn prediction system based on stream mining. In *Proceedings of the Artificial Intelligence Research and Development, 16th International Conference Catalan Association for Artificial Intelligence* (Vol. 256).
8. Huang, Y., & Kechadi, T. (2013). An effective hybrid learning system for telecommunication churn prediction. *Expert Systems with Applications, 40*(14), 5635–5647.
9. Brânduşoiu, I., Toderean, G., & Beleiu, H. (2016). Methods for churn prediction in the pre-paid mobile telecommunications industry. In *2016 International Conference on Communications (COMM)*. IEEE.
10. Hua, G. (2011). Customer relationship management based on data mining technique—Naive Bayesian classifier. In *2011 International Conference on E-Business and E-Government (ICEE)*. IEEE.
11. Huang, Y., et al. (2015). Telco churn prediction with big data. In *Proceedings of the 2015 ACM SIGMOD International Conference on Management of Data.*
12. Makhtar, M., et al. (2017). Churn classification model for local telecommunication company based on rough set theory. *Journal of Fundamental and Applied Sciences, 9*(6S), 854–868.
13. Amin, A., et al. (2016). Comparing oversampling techniques to handle the class imbalance problem: Acustomer churn prediction case study. *IEEE Access, 4*, 7940–7957 (2016)
14. Amin, A., et al. (2018). Just-in-time customer churn prediction: With and without data transformation. In *2018 EEE Congress on Evolutionary Computation (CEC)*. IEEE.
15. Rygielski, C., Wang, J.-C., & Yen, D. C. (2002). Data mining techniques for customer relationship management. *Technology in Society, 24*(4), 483–502.
16. Wu, R.-C., Chen, R.-Y., & Chen, C-C. (2005). Data mining application in customer relationship management of credit card business. In *29th Annual International Computer Software and Applications Conference (COMPSAC'05)* (Vol. 2). IEEE.
17. Sethi, I., Factors affecting consumer behaviour in telecom industry
18. De Caigny, A., Coussement, k., & De Bock, K.W. (2018). A new hybrid classification algorithm for customer churn prediction based on logistic regression and decision trees. *European Journal of Operational Research, 269*(2), 760–772 (2018)

19. https://www.qualtrics.com/experience-management/customer/customer-churn/.
20. Zhao, L., et al. (2017). K-local maximum margin feature extraction algorithm for churn prediction in telecom. *Cluster Computing, 20*(2), 1401–1409.
21. https://towardsdatascience.com/feature-selection-techniques-in-machine-learning-with-python-f24e7da3f36e.
22. Burez, J., & Van den Poel, D. (2009). Handling class imbalance in customer churn prediction. *Expert Systems with Applications, 36*(3), 4626–4636.
23. RumaPanda, A. N. (2015). A Survey of Customer Relationship Management.
24. Kim, H.-S., & Yoon, C.-H. (2004). Determinants of subscriber churn and customer loyalty in the Korean mobile telephony market. *Telecommunications Policy, 28*(9–10), 751–765.
25. Rajamohamed, R., & Manokaran, J. (2018). Improved credit card churn prediction based on rough clustering and supervised learning techniques. *Cluster Computing, 21*(1), 65–77.

The manufacturer's authorised representative in the EU is Springer Nature Customer Service Centre GmbH, Europaplatz 3, 69115 Heidelberg, Germany. If you have any concerns regarding our products, please contact ProductSafety@springernature.com

Printed and bound by CPI Group (UK) Ltd, Croydon, CR0 4YY
30/09/2026
02225306-0017